A Royal Fir

Teaching Mathematics:

Puncturing Space: A Developing Pedagogical Tool

To Yolanda Ortiz

Enjoy

Dr. Robert E. Mason IV

9/07

Robert E Mason IV

Royal Fireworks Press

Unionville, New York, 10988

Dedication

I know how much the love of my parents, Robert Emmett Mason III and Marian Williams Mason, and their passion for education have helped to sustain my enquiries. I know how much my wife, Pamela Faye, my daughter, Garnette Elizabeth, and my sister, Pamela Michele, have contributed to my sense of well-being. This book is dedicated to them.

Library of Congress Cataloging-in-Publication Data
Mason, Robert E. (Robert Emmett)
Teaching mathematics : puncturing space : a developing pedagogical tool / Dr. Robert E. Mason IV.
p. cm. – (Royal fireworks monograph on education)
ISBN 978-0-89824-352-9 (library edition : alk. paper)
– ISBN 978-0-89824-353-6 (pbk. : alk. paper)
1. Mathematics–Study and teaching (Middle school) I. Title.
QA11.2.M2678 2007
510.71'2–dc22
2007019149

Royal Fireworks Press
First Ave, PO Box 399
Unionville, NY 10988
845 726 4444
FAX: 845 726 3824
email: mail@rfwp.com
website: rfwp.com

ISBN: 978-0-89824-352-9 Library Binding
ISBN: 978-0-89824-353-6 Paper

Printed and bound in the United States of America by American citizens using using soy-based inks on acid-free paper and environmentally-friendly cover coatings at the Royal Fireworks Printing Company of Unionville, New York.

Table of Contents

Acknowledgements

The people who have contributed so much to my sustained commitment to this educational inquiry are acknowledged in my publication. I feel sure you know how much I have valued your company and help in moving on my enquiries. Thank you.

As I complete this phase of my professional practice I am fortunate to have worked with colleagues at the Dalton School; Dr. Gardner P. Dunnan, Dr. Frank Moretti, Mr. Peter Sommer, and Mrs. Kamian Allen, Mrs. Nina Goldberg, and Ms. Debbie Coons. A special thanks to Dr. Raphael Kasper, for without his patience and editorial skills this manuscript would not have become a finished product.

Foreword
by
Dr. Kenneth Offit

Teaching Mathematics: Puncturing Space: a Developing Pedagogical Tool by Dr. Robert Emmett Mason IV, does not fit an easy description. It is part authoritative teaching handbook, part textbook, and part philosophical discourse from a master pedagogue with thirty years teaching experience. It is a marvelous resource for all teachers of mathematics and the natural sciences.

From his very first teaching job in 1973 at a school for gifted children in New York, Dr. Robert Mason has been taking careful notes and developing a philosophy of teaching that he shares in this elegant and incisive volume. The intellectual framework for his teaching approach is modified from Jack Whitehead's "Living Educational Theory of Teachers," in which teaching becomes an iterative process of self-discovery on the way to unlocking the intellectual capacities of students. Robert Mason develops a model in which the teacher moves from instructionalist to constructivist educator, constantly modifying and improving didactic approaches in response to the dynamics of a classroom.

The first chapter of the text is a highly personal and riveting account of Dr. Mason's earliest experiences and development as a mathematics teacher. It chronicles his transformation from a novice teacher clinging to the Teacher's Edition of a standard textbook to a skilled craftsman who challenges and is challenged by the dynamic interactions of the classroom.

This personal theme is continued as Dr. Mason applies classical models of learning theory such as those by Piaget and Vygotsky to further define his pedagogical approach. The book includes a fascinating meditation on the most basic notions of consciousness that are a prerequisite to understanding the basis of the classical admonition of the teacher: "please give me your undivided attention." Abstract sections touch on a wide range of themes, including theories of neuronal group selection and brain physiology and attempts to define the "boundedness of the contents of mental activity." Such biological and psychological foundations, Mason argues, are critical for the teacher to understand and improve strategies of communication of mathematical concepts.

Teaching Mathematics represents that rare synthesis of cognitive/learning theory and a practical annotated curriculum. Dr Mason is the creator of *Euclid II*, as well as the prior annotation of *Geometers Sketch Pad* that has been widely utilized in classrooms for years. The two new curricula proposed in this volume can stand by themselves and provide exciting tools for teachers seeking to provide novel course outlines for mathematic students in either public or private educational settings. What is unique about this monograph is the juxtaposition of a highly innovative, detailed mathematics curriculum with the philosophical construct from which it emerged.

Having witnessed Dr. Mason as a teacher, I can also testify that he is an expert practitioner of his pedagogical theories. Dr. Mason's sensitivities to cultural and gender differences among students have nothing to do with "political correctness," but rather stem from his heartfelt philosophical commitment to the individualized didactic approach to education as articulated in the opening chapters of this book. While there is perhaps no substitute for witnessing Dr. Mason's teaching in the classroom, this volume provides an example of an eloquent and highly readable "concept proxy" devised by a gifted educator.

Kenneth Offit, M.D., M.P.H.
Professor, Medicine and Public Health
Cornell University Medical College
Chief, Clinical Genetics Service.
Memorial Sloan-Kettering Cancer Center

The Importance of Teacher Experience in the Classroom

Those who can, do. Those who can't, teach.

We have all heard this statement expressed by intelligent individuals, individuals who themselves are successful in fields other than teaching. I find the sentiment most insulting and misleading—insulting because it demeans and devalues the efforts and achievements of past and present teachers, scholars, philosophers, and theorists in the field of education, and misleading because it carries the implication that intelligent people choose not to teach because teaching does not require intelligent "doing." Furthermore, the statement can only be based upon the personal experience of the speaker, and that experience is, of necessity, based only on interactions with a small and particular set of teachers. The sentiment expressed bears only on the substantive personal knowledge those teachers possess. Yet, there are broader intellectual components of teaching that are often overlooked or ignored because they are rarely discussed with students or, indeed with those not intimately involved in public or private education. For instance, many teachers spend countless hours thinking about the learner and activities that promote "learning," in an effort to determine how best, or most effectively, to stimulate the mind of students to store, retrieve, analyze, synthesize, and evaluate information.

The first step in *intelligent doing* occurs when teachers analyze and assess various *theories* of learning. As teachers, we sort through competing theories, then spend years translating educational theories into practical applications that seek to "unlock the intellectual potential of the students." When I say we seek to "unlock students' mathematical potential," I mean that students come to a class with a store of knowledge and their potential—yet are often unaware that they possess this knowledge or that their existing knowledge of mathematics is part of a larger system. Enabling students both to recognize the knowledge that they possess and to relate that knowledge to new information should be the goal of a curriculum. An effective curriculum should provide situations in which students experience this recognition and internalize these relationships, and the ones I describe in this book provide a working example. I call this kind of thinking and curriculum development "***intelligent doing***."

I think that it is fair to assume that most non-teachers would at least acknowledge the fact that teaching is done in a social context that is complex. And, they would agree that the intellectual potential that teachers strive to unlock is sometimes easily accessible and sometimes not. Moreover, most would agree that some students possess intellectual abilities that require a qualitatively different kind of "unlocking" than those whose potential is not easily accessed.

How, when, and whether teachers get to the stage in their teaching where they actually think about teaching and learning is still not well understood. I believe that some teachers go through what I call a pedagogical transformation that occurs as a result of direct as well as indirect interactions with the cultural framework of their schools, that is, with the implicit and explicit values, traditions, beliefs, patterns of behavior, and assumptions that are held by the various constituencies of the community. It is within this context that teachers develop their approaches to teaching and determine the degree to which they satisfied, or dissatisfied, with their pedagogical, or instructional practices. It is within this context that the "doing" of teachers is intelligent.

Part of the problem lies in the fact that teachers have failed to share the details of their intelligent doing with a larger audience. And, because of this, non-teachers have formed an opinion of teaching that is based on incomplete information and thus they lack an appreciation of the depth and complexities of our work. Their opinion is "boundedly rational:" rational because it is based on the experience of at least twelve years of formal education, and bounded because their teachers are but a small sampling from the total population of K-12 teachers in the country.

Let's think about what I've said. When we are asked what we do for a living, teachers generally reply by naming the subject they teach and to whom—end of conversation. For the most part we don't discuss the *details* of our intelligent doing or our efforts to *unlock the intellectual capacities of our students*. If we are to influence the opinions of those outside our profession, it is essential that we communicate the scope and extent of our efforts to the larger population so that we can bridge the gap between what non-teachers think they know about our work and what we actually do. One way this can be accomplished is for teachers to begin to contribute publicly to intellectual explorations—inquiries of the theory and practice through action research.

For instance, classroom teachers can and should participate in general discussions about theory and practice in a manner similar to the way a professional race-car driver relates to his team of expert mechanics. The experts attempt to design and create innovations that will improve the performance of the racing car. Professional drivers in turn give feedback in terms of their interpretations, the meaning that they attach to their interpretations, and the evidence of how they affect the performance of the automobile. The experts then evaluate their theories and designs in light of this feedback and, when appropriate, modify their theories. The cycle is clear and effective for automobile racing. But, while teachers are often asked to adopt new theories of learning and pedagogical practices, only infrequently does input from them flow back to the original sources in the form of an articulated analysis and synthesis of the theories. Such syntheses, if they were developed, could be in the form of what Jack Whitehead calls, "living educational theories (LET)" of teachers (Whitehead 1984).

If teachers were to participate in the process (much the way racing car drivers do) they would be, in principle, researchers in a practice context. They would not merely apply established theory and techniques but would be central figures in the analysis and assessment of the theory and technique. Their involvement would feed back to the developers of the theory and technique so that the original ideas could be altered and adjusted by incorporating the experience of actual practice.

That is, when a teacher reflects-in-action, he/she becomes a researcher in the practice context. He/she is not dependent on the categories of established theory and technique but constructs a new theory of the unique case. Living educational theories explain events by setting forth propositions from which these events may be inferred, a predictive theory sets forth propositions from which inferences about future events may be made, and a theory of control describes the conditions under which events of a certain kind may be made to occur. In each case, the theory has an "if...then...." form (Schon 1983). These contributions would give the original sources insight into their work in terms of how to interpret them and use them in practice.

The importance of teachers' stories of practice as examples of their intelligent doing.

Cochrane-Smith & Lytle (1990), and Carter (1993) contend that teachers' stories of practice are largely an untapped source of information about teaching. Group effort has the potential to help teachers individually as well as the profession as a whole by drawing attention to teachers' stories of practice. They also make a compelling case for why teachers should engage in group inquiries into teaching, stating that missing from the knowledge base for teaching are the voices of teachers themselves, questions teachers ask, and the interpretive frames teachers use to understand and improve their own classroom practice. Florio-Ruane (1991) notes in his research the traditional forms of educational research are largely done by university professors and often do not capture the perspectives and experiences of the classroom teacher.

Polkinghorne (1988) has theorized that humans tell stories in order to make meaning from experience. His research into what kind of knowledge teachers use in practice shows that teachers use stories or "narrative ways of knowing." Closely related to Polkinghorne's research is the work of Connelly and Clandinin (1990). They suggest the stories teachers tell about their work are key to gaining insights into what teachers know about teaching and how they come to know. Connelly and Clandinin's extensive collaborative research with teachers (Clandinin 1985; Connelly & Clandinin, 1988; Clandinin, Davies, Hogan & Kennard, 1993; Clandinin and Connelly, 1995) indicates that these stories about practice tend to benefit others, and equally important to observe is that teachers themselves learn in the process of telling, retelling and interpreting their stories. Elbaz (1983) concluded from his research that teachers' knowledge is personal, embodied, narrative, autobiographical and relational. These researchers stress the importance for teachers of understanding their own teaching narratives—their narrative ways of knowing.

Witherell & Noddings, (1991) are strong advocates for recognition of the importance of narrative and dialogue in education. The stories that teachers tell about their own practice are examples of their "intelligent doing" because their voices add to the research base on teaching. The stories teachers tell do not reveal teaching and learning as straightforward but as involving a complex interaction with students and content. Coles'(1989) and Paley's (1979) published accounts of their teaching provide further insight into the use of story as a teaching strategy and the importance of story in their teaching lives.

There is mounting support for, and acknowledgement of, the significance of what classroom teachers have to tell about their practices in the research arena. In time, the stories that teachers tell about their practice will improve and will be viewed as a valuable and useful source of information.

Teaching Mathematics is the story of my personal journey in search of a practical link between my understanding of the Piagetian stages of cognitive development, Vygotsky's social constructivist theory of mediated learning through cultural artifacts, and my practices of teaching mathematics to students.

While teaching has intense idiosyncratic aspects, it also has general applicable aspects. This book can't transmit the idiosyncratic aspects of Robert Emmett Mason's experience, but it can illuminate how one teacher went about developing his pedagogy. This example may prove useful to others as they undergo their path to "intellectual doing."

I view teaching as a continuous competition between the internal stimuli of students and the external stimuli that would attract their attention. **I will show how the act of teaching can be viewed as "puncturing space" with pedagogical objects, though this will take some extended explanations.**

I will then propose the following conjecture: Pedagogical space is punctured with objects for the sole purpose of winning the competition for the attention of students that is ongoing between the internal stimuli of the learner and external stimuli from the environment. Such a practical conceptualization of the classical environment in which teaching occurs provides a framework in which teachers can think about their pedagogical practices, their role as the teacher, and their view of the role of the student.

In my experience, a bridge can be formed between an understanding of theories of learning and the practice of teaching through the general notion of what I call, "functional closure," or the tendency of human beings to "bound" several aspects of their psychological and physiological realities.

Throughout this monograph, I emphasize the theme that teachers can make significant contributions to the field of education in intelligent and creative ways. Because I believe strongly that this is not merely a possibility but, in fact, an achievable goal, I illustrate this point with a detailed discussion of my own transformation from an instructionalist teacher into a constructivist educator.

This book itself seeks to be an example of what I've discussed above. It is my contribution to the field of education in the form of an analysis and synthesis of ideas that

I have been taught in graduate school, gained by personal reading, and developed through experience teaching mathematics to elementary and middle school students. It is an example of the kind of intelligent doing that is possible from classroom teachers.

Because my ideas evolved from, and are the results of, a search for understanding illuminated by my practice in teaching math, the story begins with a description of my thoughts and concerns during the beginning years of my career as an elementary school teacher.

Chapter One is a discussion of how the workplace fashioned my initial beliefs, values, assumptions, and perceptions about teaching and learning and how, during the first three years of teaching, I found, *enlightenment* from my interactions with students who challenged those starting points.

Chapter Two describes the period in my self-development when I began searching for meaning in my teaching. I describe, discuss, and interpret the research results on teaching, learning and related fields. One could characterize this period as an *awakening*.

Chapter Three turns to a third period of my pedagogical transformation, a period of unconscious intellectual work, or *incubation*.

Chapter Four continues the detailed description of my unconscious intellectual work and then derives the notion of teaching as **Puncturing Space with Pedagogical Objects.** This is the period I call, "resolution and closure."

Chapter Five, referred to as *more thinking*, introduces a new construct that I call **"conceptual proxies"** and describes how they are paths to what can be thought of as conceptual identity.

Chapter Six, entitled, "Vygotsky, Piaget and The Theory of Neural Group Selection;" is a consideration of a fascinating theoretical debate about the work of two significant men, Lev Semenovich Vygotsky and Jean Piaget. In this chapter, I discuss how the conceptual framework of "puncturing space with pedagogical objects" and the construct "concept proxies" provide a conduit for linking the perspectives of the two bodies of work. Vygotsky's concept of the "zone of proximal development" is operationally defined and then used in conjunction with the concept of puncturing space to re-establish on-task behavior of students. His notion of "culturally based artifacts" is illustrated through the construct of "concept proxies."

Not included in this book, but available from Royal Fireworks Press are detailed descriptions of two curricula that were designed with the notions of puncturing space at its core. The first is entitled *Euclid II: Advanced Geometry Topics on Geometer's Sketchpad* and is the second geometry curriculum in this series. This non-traditional approach to learning geometry invites students to ponder, question, speculate, and discover for themselves the theorems, and definitions of Transformational Geometry. Students use Geometer's Sketchpad as their electronic *tabula rasa* on which they illustrate geometric proofs.

Each mini lesson is based on Benjamin Bloom's *Taxonomy of Educational Objectives in the Cognitive Domain* and the standards for teaching mathematics to middle school students developed by the National Council of Teachers of Mathematics (NCTM). Students gain a clear understanding of the theorems when they illustrate them on Geometer's Sketchpad. They apply their knowledge of specific concepts through practice worksheets and through problems designed to tap the higher order cognitive skills of analyzing, evaluating and creating. These thinking skills are also addressed through the writing component of the program when students write comparison and contrast essays on two, or more concepts.

The second curriculum entitled, *An Introduction to Experimental Research Skills for the Young Scientist,* was developed for motivated fifth through eighth graders. It is designed to enrich and extend the traditional science program for children who excel, and whose abilities and talents are not being developed adequately in the regular classroom. The program is also designed to develop the student's capacities for problem solving and critical thinking in science.

Chapter 1

Retrospection: My Humble Beginning

My first teaching job began immediately after I graduated from college in 1973. It resulted from an unexpected offer; I had been planning to teach at a public school in Brooklyn, New York but was instead recruited by Hunter College Elementary School for the Intellectually Gifted in Manhattan to teach mathematics to fourth, fifth, and sixth graders. With no experience and armed only with a textbook understanding of teaching and learning, I entered the world of teaching in September 1973.

The last two weeks of August and the first two weeks of September were filled with anxiety, sleepless nights, and long hours of planning and writing a curriculum. I decorated my classroom with colorful paper and child-centered posters—nothing particularly creative or original but designed, I hoped, to make the room feel warm and inviting.

As opening day approached, my anxiety increased. In fact, I did not sleep at all the night before. On my way to work, I asked myself a myriad of questions: How do I begin? What should I do first? And on and on. The principal and several colleagues dropped by my classroom and offered supportive comments. It was clear from the suggestions that all of the teachers approached the opening day of school differently. Some suggested that I should establish a business-like tone and make clear to the students that the classroom atmosphere would be focused on work, work, and more work. Others suggested that I establish a sense of community by pursuing a nurturing approach. In the end, I adopted an approach that included a little of both—an approach that established a set of daily, weekly, and monthly routines.

Those routines addressed such recurring activities as orchestrating classroom events, managing transitions between activities, dealing with conflicts, communicating with colleagues and parents, keeping track of textbooks, attending to record-keeping, ordering class supplies, writing lesson plans, creating tests, and writing report cards.

One reason I began to understand the importance of organization early in my teaching career was because the principal of the school initiated a very successful mentoring program, one emphasis of which was the importance of established routines and the necessity of both the appearance and reality of a sense of order in the classroom. Colleagues, who believed that I had the right "stuff" eventually to transform into a "good" teacher, volunteered to meet with me after school two days each week. The informal meetings were of great value to me, because I certainly had a host of unanswered questions. During the meetings my colleagues shared information about three significant constituencies—parents, teachers, and students. They also helped me plan and organize my work.

The meetings also illuminated the values, beliefs, established practices, and the many do's and don'ts that were particular to the culture of the school. There were, after all, things about this elementary school that only someone intimately involved with it could possibly know. Such as, what model of education the school followed, which teachers had the greatest influence in the organization, and which parents were most actively involved in school affairs. My mentors helped, as well, when I, inadvertently, but inevitably "stepped" on another colleague's toes, or when I disagreed, or challenged, a senior staff member. Like all new teachers, I was perceived as a novice who had no experience and hence could not have an "informed" opinion. Notwithstanding that, most of my colleagues willingly shared their books, lesson plans, and teaching materials with me. But, helpful as they were, the one area in which they could not assist with was the actual teaching of mathematics.

If I had to use a single word to characterize my first year of teaching it would be "overwhelming." I had set as my first priority learning to manage effectively the dynamic environment of the school and classroom and achieving this one goal occupied all of my time. Because getting to this goal was so all-consuming and important, I found it virtually impossible to inspire myself to be innovative in my work.

During my first year I viewed teaching as merely the dissemination of information. Because I knew the mathematics that I taught, I simply shared my understanding with the students through lectures and demonstrations. As a result, my initial view of my role was that I had been hired to impart a small body of knowledge to the students, and my initial view of the students was that they were the recipients of that knowledge.

Curriculum Planning was directly related to my view of teaching and learning. I cannot imagine how I could have gotten through the first year without access to a very good math textbook series—more importantly, to the teacher's edition. The teacher's edition of the text provided a scope and sequence, several mini lessons that preceded the practice sections of the lesson plans, test worksheets, retest worksheets, re-teaching worksheets, and enrichment activities with written explanations of each concept and skill. I didn't even have to solve the mathematics problems that appeared at the end of each chapter of the text, nor did I have to create tests for each unit, mid-term exams, and final examination because the authors of the series provided them for me. The availability of these resources allowed me to teach a new concept each math period, and to have students practice the skills during class time and for homework. All that I needed to do was select the appropriate worksheets, run off copies, hand them out, and correct them after they were returned to me.

My lesson plans were directly modeled upon those provided in the text. Each lesson began with an activity around which I filled in the necessary parts. Typically, the formal written plan included several behavioral objectives, the material needed, a description of how to introduce the lesson, a set of important questions, a closing activity, and a follow-up section.

Assessment of the academic performance of my students was very much related to the manner in which mathematics was taught. The cycle went like this: I taught a new skill—for instance, multiplication of decimals. Students practiced the skill by solving several problems from the textbook, or practice worksheets. At a later date, students were tested to determine their level of mastery.

When I sensed that students needed more enrichment in math they were assigned the "harder" problems from the text. As more students came to need more challenging work, I accelerated the math curriculum. That is, fourth graders did fifth grade math, and fifth graders did sixth grade work, and so on.

In that first year, the textbook-based approach proved to be an effective way for me to provide a well-structured, logical, and efficient mathematics curriculum to my students. It also freed time for me to work on establishing patterns of classroom behavior, routines, and practices as well as to learn how to coordinate elements of the math curriculum, understand the formal and informal social structures of the school, and establish a fair and ethical set of rules. I learned to manage the dynamics of my specific class through trial and error and for the most part in isolation. But, although the textbook and teacher's edition allowed me to "survive" the first year, I had a growing sense that I was not yet doing all that I could to help the students truly learn mathematics. I resolved to dedicate at least some of my effort and attention in the subsequent years to improving my teaching and enhancing the learning of my students.

My second and third years began on a more positive note. Having gotten through my first year, I began years two and three with a modest history and experience. Established routines and classroom practices were easier to implement, and I had begun to anticipate the patterns of behavior of the school and of my students. I knew, for instance, that new students generally needed extensive review of skills after two months away from academic work; seasonal changes affected everyone's moods; students were generally less receptive to new ideas before vacations and holidays; and, as spring approached, students become energized in many ways—some productive and some not.

School-wide expectations and demands also affected teachers' moods and attitudes. In some public elementary schools, teachers prepare and write between thirty to forty report cards three times each year. And at the end of this flurry of activity, emotions run wild and tempers flair; report writing is followed by parent-teacher conferences. A similar emotionally draining and psychologically intensive period occurs towards the closing of school when record cards, final report cards, and final attendance records are all due. The last week of school is also generally used to prepare the rooms for summer cleaning by organizing books, personal files, and teaching supplies. Some teachers use this time to order new materials for the next school year, to make preliminary arrangements for school-related trips, and to sit on planning committees.

The culture of the school influences every aspect of the work environment. Both internal and external influences act to shape, and mold the teachers' view of the learning environment, how they respond to children and how they teach. Success, or survival, depends on how well new teachers observe, acknowledge, and manage these ubiquitous elements of their work place, so I focused on understanding the interaction of these influences on my work during my second and third years of teaching.

Chapter 2

Awakening: I Began To See Things In A Different Light

After three years of teaching, I had become a fairly effective instructionalist teacher of mathematics in the behaviorist tradition. Instructionalist teaching views learning as the strengthening of bonds between stimuli and correct responses, and hold that learners are passive receivers of knowledge, vessels into which information is poured.

Why was it so easy for me and, I might add, for so many others to use this model? I believe there are two explanations. First, we live in a world where verbal communication dominates. From the time we were toddlers through our formal schooling, adults shared their values, beliefs, and knowledge with us by telling, explaining, and demonstration. Our experience shows us that direct communication can be an efficient method of transmitting information. Second, as mentioned earlier, in the early years I was principally concerned with establishing routine patterns of behavior for both my students and myself and with learning and adopting the political and social aspects of the school. I, therefore, resorted to the mode of teaching that was the easiest to replicate and employ. The instructionalist model of teaching was the style used predominantly by most of my colleagues, many of whom I admired and respected. These teachers had masterfully constructed a form of teaching that combined the use of a textbook with their abilities to tell, explain, and demonstrate as a way to infuse students with information. The instructionalist model was necessary for me to survive the early years, and it made sense to me because it provided a structure in which I was able to control competing elements of a dynamic and interactive environment. It represented the first practical example of "intelligent doing."

I characterize my fourth and fifth years as an "awakening" when I was ready for the next step in a changing perspective of my methods, or style of teaching. And so I began my transformation from an instructionalist (subordination of learning to teaching) teacher into a multiple perspective teacher in which constructivist (subordination of teaching to learning) tendencies dominated my classroom practices. The change in perspective resulted from an acknowledgement that my students would often respond to instruction in qualitatively different ways when I deviated from a mode of teaching which emphasized telling and showing (this statement will become clearer as I develop my story).

During the second and third years, I had accumulated math materials of all kinds and engraved the scope and sequence of the math textbook in memory. I could have replicated by rote the same lesson next month, or next year, without hesitation. I merely had to re-construct a "pedagogical script" in my mind. This developing ability was a fascinating new experience for me. Previously, I had used written lesson plans as a way of organizing lessons, often referring to them as I implemented the lessons. But now I rarely referred to my lesson plan. Recalling my "pedagogical scripts" manifested some residual benefits. I noticed that I was now becoming more sensitive to students' verbal responses as I taught. I was now orchestrating learning by encouraging students to refine their questions, infer from them, and finally evaluate the results. The most significant results of a heightened sensitivity to the students was an increased awareness that I was competing with the internal mental activities of my students and the external stimuli from the environment for the student's attention.

I believe we use the phrase "selective attention" in terms of our observations of students during classroom activities. We look for signs of on-task behavior—are students listening carefully to the lecture, are they following the logical development of a demonstration, and are they fully participating in their group activity? Teachers develop strategies which allow them to make spot checks to determine if students are paying attention during learning situations. One strategy often used by teachers is questioning at various points during the lecture or demonstration. When students fail to answer questions appropriately, the teacher may conclude that the student was not paying attention. Although there are several reasons and perhaps a number of unknown reasons as well that could equally explain why the student was not paying attention, the point here is that the teachers compete with them to achieve educational objectives.

Very narrowly defined and context specific to teaching, selective attention is the sometimes controlled response to incoming external stimuli and the processing of internal stimuli to integrate multiple sensory inputs. When students attend to a particular learning activity, they are said to be motivated to learn.

Lepper (Lepper 1988) suggest that there are two types of motivation. Intrinsic motivation is learning for its own sake, for the enjoyment it provides, or the feelings of accomplishment it evokes.[1] Research suggests that intrinsically motivated students tended to employ strategies that demand more effort and processed information more deeply (Lepper). Extrinsic motivation is learning in order to obtain some reward external to the activity itself, such as grades, or my approval.[2]

Although every educational activity cannot, and perhaps should not, be intrinsically motivating, there are several potential benefits when teachers can capitalize on existing intrinsic motivation.

Whereas this interpretation was undoubtedly correct for some students, it did not explain why the phenomenon oc-

[1] Research suggests that intrinsically motivated students tended to employ strategies that demand more effort and processed information more deeply (Lepper).

[2] Extrinsically oriented students are inclined to put forth the minimal amount of effort necessary to get the maximal reward (Lepper).

On the same topic, Condry and Chambers (1978) found that when students were confronted with complex intellectual tasks, those with an intrinsic orientation used more logical information-gathering and decision-making strategies than did students who were extrinsically oriented. Students with an intrinsic orientation also tend to prefer tasks that are moderately challenging, whereas extrinsically oriented students gravitate toward tasks that are low in degree of difficulty.

curred in approximately 60% of the students in my math classes each year.

The Apple Metaphor

I significantly changed my methodology and classroom practices as a result of an experience that occurred while I was teaching at Hunter College Elementary School. One day I had an unannounced visit by thirty middle and high school teachers from England. I was trying an experimental lesson that day to test the effectiveness of using metaphors to teach symbolic manipulations to a group of sixth graders. I wanted to determine if students would see the similarities between the "undressing of an apple to reveal the number of seeds in its core," and "the transformations used to isolate an unknown in an equation." As the thirty visitors walked into my class I had written on the chalkboard the question: "How is undressing of an apple like solving a linear equation?"

A drawing of an apple also appeared on the chalkboard. As the lesson developed, I sensed a change in the classroom environment. The manner in which I had approached this lesson had captivated my audience. Students were tracking every move I made and appeared to hang on each word I spoke. At first, students and teachers alike were curious about where I was leading them—from their nods

and smiles I soon realized that their understanding and acceptance of the metaphor was converging towards the understanding and acceptance of the metaphor that I had hoped would result from this lesson. The lesson went well but, more importantly, I now had a mental picture of a lesson that had increased the number of students who paid attention and who were at least curious, if not motivated, to learn the material.

What did I do differently in my teaching of the metaphor lesson[3] which set it apart from how I usually practiced teaching? It was the first time that I had used metaphorical teaching strategies to teach mathematical concepts. Although the lesson involved direct instructions, my students were allowed to discuss the concepts and challenge the statements of their peers and the teacher. I also gained new insight into the students as problem solvers and was able to observe which students were the logical and creative thinkers and which students had a range of knowledge of mathematics and could use that knowledge to explain or solve problems.

When I applied the notion of "competition for attention" in my analysis of the metaphor lesson it became clear that my "teaching" (a potential extrinsic motivator) had engaged a larger number of students.

[3] The Metaphorical Teaching Strategy based on the pioneering work of W.J.J. Gordon (1961), students engage in a metaphoric journey using three different types of analogies: direct analogies, personal analogies, and compressed conflicts. The strategy is designed to help students make the strange familiar or the familiar strange.

Had I won the competition between the internal mental activities of my students and the external stimuli for their attention?

According to Jere Brophy (1987), motivation to learn is a competence acquired "through general experience but stimulated most directly through modeling, communication of expectations, and direct instruction." Given this description of motivation, I would say, yes, I had.

For several days thereafter, I thought about the students' reaction to the lesson, and the more I thought about the lesson, the more I yearned for a deeper understanding of it—an understanding that was based on accepted and established theories of learning to which I could attach my own interpretation and experience.

Searching through graduate school papers on learning theories, I found my lecture notes on cognitive constructivism as viewed by Jean Piaget, and social constructivism by L. Vygotsky. I also found my notes on motivation and learning. Integrating my understanding of both cognitive and social constructivism with my experience teaching the "apple" metaphor, I realized that I had observed, first hand, students in the Piagetian cognitive stage of concrete operational thinking, that is, students who were capable of logical thinking when it is attached to concrete examples. Apparently the "apple" metaphor assisted these students in their "meaning-making" because it was a model that provided a useful representation of knowledge that was needed to understand fully the algebraic manipulations and concepts. This interpretation is consistent with the findings of Gage

& Berliner (1992) who stated that the use of models as learning aides haas two primary benefits. First, models provide "accurate and useful representations of knowledge that is needed when solving problems in some particular domain" (p. 314). Second, a model makes the process of understanding a domain of knowledge easier because it is a visual expression of the topic.[4] Alesandrini (1981) came to similar conclusions when he studied different pictorial-verbal strategies for learning:

> Research on the effectiveness of pictorial learning strategies indicates that learning is improved when pictures supplement verbal materials, when learners draw their own pictures while studying, and when learners are asked to generate mental pictures while reading or studying...the factor of gender was also included in the analysis due to its observed (although unexpected) effect (pp. 358, 363).

By identifying each concrete element of the apple (a picture) with a symbolic counterpart, these students were able to appreciate and understand the formal algebraic manipulations. Moreover, I had finally interacted with these students within what Vygotsky calls, their "zones of proximal development, or ZPD", because the use of the metaphor facilitated the possible link between the apple represented as figural information and the equation as symbolic information through language. Therefore, the zone (a psychological space) in which this instruction occurred fell between what students could do for themselves and what they could do with assistance from me.[5]

[4] Gage and Berliner also found that that students who study models before a lecture may recall as much as 57% more on questions concerning conceptual information than students who receive instruction without the advantage of seeing and discussing models.

[5] There is no single ZPD for individuals because the zone varies with culture, society, and experience (Tharpe & Gallimore, 1988). Vygotsky (1962) claimed

During the subsequent weeks I replayed the lesson repeatedly in my head. Interesting, and perhaps tellingly, conversations with my colleagues failed in a large part because we lacked a common vocabulary of terms, definition, and tacit theories. Moreover, as my colleagues had not read the current research and theory on teaching and learning, they were unable to assist me with the analysis of the lesson.

Most of our more successful discussions about teaching centered on how we felt about a particular child and upon our interpretations of these feelings. We openly and willingly shared curriculum materials, lessons, and insights about the potential of students that were based on our evaluations of them. But only infrequently did our discussions focus on treatment of the elements of our practice and the manner in which they elicit qualitatively different patterns of behavior from students. We rarely discussed specific pedagogical issues in terms of the intellectual practice of students within the classroom community. For instance, we rarely examined whether students who appear to think more clearly when dealing with concrete objects make meaning differently from students characterized as abstract thinkers. I was essentially on my own to seek answers to my questions and concerns.

So, I began, in the sixth and seventh years, to reread the literature on teaching, learning, brain research, decision

that the larger the zone, the better students will learn in school. For a ZPD to be created, there must be a joint activity that creates a context for student and expert interaction. The expert may then use multiple instructional strategies (Tharpe & Gallimore, 1988). Social interaction is important because the expert can model the appropriate solution, assist in finding the solution, and monitor the student's progress (Tharpe & Gallimore, 1988).

making, and thinking, seeking information that would help me better understand, and subsequently improve, my methods of teaching mathematics to children. It quickly became obvious that there were several competing, or at least not completely consistent, points of view. Some theories of learning assume that if we know how learning occurs, we thereby know how to teach. For example, it is generally assumed that if we know how individuals solve problems, then we know how to teach by the problem method. Other theories attempt to apply philosophical or scientific knowledge directly to teaching. Still others always speak of teaching and learning together as the teaching-learning process. Lamon and Lesh (1992) describe the state of affairs as:

> Behavioral psychology (based on factual and procedural rules) has given way to cognitive psychology (based on models for making sense of real-life experiences), and technology-based tools have radically expanded the kinds of situations in which mathematics is useful, while simultaneously increasing the kinds of mathematics that are useful and the kinds of people who use mathematics on a daily basis. (p. 18-19)

Although aspects of all views made sense to me, I found it very difficult to apply many of them. And I worried that if I could not specifically match my teaching practices to these wonderful ideas, then would I be unable to explain to myself and to others what it is that I do each day in my math classes?

Bridging a personal understanding between the theories of others and my own practical pedagogy required a process of synthesis. I searched for a conceptual framework of the classroom (the social context of teaching and physical objects) that would relate three aspects below:

1 Teaching competes for students' attention with internal mental activities and external stimuli from the environment.
2 Logical thinking occurs when it is attached to concrete examples[6]
3 There is a zone of proximal development.[7]

What originally began in my conscious mind had now moved to my subconscious. I believe that is a way the mind processes information in creative ways and where it is safe to take risks. Even when the problem was removed from my direct attention, my subconscious mind was working on it. Since ideas and fantasies about possible solutions surfaced only transitorily during this period, I took to writing down my dreams so that eventually they would be clarified and organized. During the next two years' teaching,

[6] The child is able to solve concrete problems in a logical way, but s/he is still unable to think hypothetically, which is more an abstract form of thinking. The thinking focuses on describing what an object is but not on what an object might be. His/her thinking has to depend on concrete objects, that is, the symbols in his/her mind have equivalences in the concrete form. But when the child is asked to think about something that has no concrete equivalence, like the concepts of democracy, values, s/he has trouble in operating on them. Some past experience with physical objects is needed for imagination. Education of the concrete operational child requires the use of concrete objects, examples, pictures and hands-on practices to help him/her understand certain concepts.

[7] One essential tenet in Vygotsky's theory is the notion of the existence of what he called the zone of proximal development. Zone of proximal development is the difference between the child's capacity to solve problems on his own, and his capacity to solve them with assistance. In other words, the actual developmental level refers to all the functions and activities that a child can perform on his own, independently without the help of anyone else. On the other hand, the zone of proximal development includes all the functions and activities that a child or a learner can perform only with the assistance of someone else. The person who intervenes in this scaffolding process could be an adult (parent, teacher, caretaker, language instructor) or another peer who has already mastered that particular function.

my thoughts, ideas, and concepts had begun to coalesce in a coherent conceptual framework for my teaching philosophy and practice.

That framework now is used to guide and inform the work I do with children. I have chosen to present this conceptual framework to the reader by describing as much of the journey from the inception to closure as I can reconstruct.

Chapter 3

Incubation: Thinking in the background

I made a conscious decision not only to improve my teaching, but to understand my world from my own point of view. My attempts to explain what I observed, interpreted, and sought meaning in were and are part of a real attempt to improve my professional practice; they suggest that I yearn to attach meaning to my own pedagogical development.

The story that I am about to share with you may at first appear somewhat strange because most of us do not pay much attention to our dreams.[1] Dreams are products of our subconscious mind and are often extremely complex and ambiguous. Many cultures believe that our dreams represent unconscious problem solving undertaken when the mind is most relaxed and receptive to creative and divergent thinking.

[1] The term "lucid dreaming" refers to dreaming while knowing that you are dreaming. The "lucid" part refers to the clarity of consciousness rather than the vividness of the dream. It generally happens when you realize during the course of a dream that you are dreaming, perhaps because something weird occurs. Most people who remember their dreams have experienced this at some time, often waking up immediately after the realization. However, it is possible to continue in the dream while remaining fully aware that you are dreaming.

The dream that I will describe in the next few pages of this text was not a one-time occurrence; it was a dream that had been recurring for a number of years.

The Dream

My dream has a certain abstract quality and begins with two spheres that encompass "chunks" of fireworks-like activity linked together by several paths, chains, or trails in an open-ended, ongoing, incomplete activity structure within an amorphous space. A third inactive sphere lies dormant in the distance. The activity within the first sphere appears to demonstrate some order or pattern. At times, the internal activity of the spheres crosses the boundaries of their surface area space. At other times the activity outside the spheres penetrates the surface on its way toward the center of the spheres. The closest physical model to my dream that I have encountered is a toy that is sold in major department stores. The toy consist of a transparent sphere that is resting on a rectangular base. When the sphere is touched, the center of the sphere generates what appear to be random streams of electricity. Objects near the surface of the sphere (one's fingers, for example) attract the streams of electricity.

At first, the dream made very little sense to me, but eventually the elements of the dream were replaced with more recognizable and comprehensible objects. The dream itself began to develop a coherent form or structure. In the transformed dream, the activity that danced within and between spheres could be interpreted as streams of thought patterns, patterns that varied in length, density, and velocity.

My mind had begun to associate the internal activities of the spheres with cortical activities in my own brain. As the transformation of the dream continued, the two active spheres were replaced with images of a teacher and a student, and the inactive sphere in the distance changed into a concrete set of objects—a set of black and white chips. The images did not interact at first, but the space between them became filled with other objects (that took the form of things one would expect to find in a classroom). The objects filled the space and gave it "density." The objects were recognizable, but as my mind constructed a new conceptual system they became redefined and took on new meanings.

The Conceptual Framework

At the outset, students and teachers possess information, all that they have accumulated up to the beginning of this hypothetical interaction. We may not know the precise quality or quantity of the information possessed by either individual, but we can assume that the knowledge is both stored and is retrievable. We can refer to the acts of storing and retrieving of information as examples of the many forms of mental activity. We assume that both individuals have the capacity to store and transmit information and that their states of information change as they construct meaning of what is perceived in the environment. To understand the hypothetical observation in terms of my description of teaching (the competing with the internal mental activities of my students and the external stimuli from the environment for the students' attention), I needed an accepted notation and vocabulary that would allow me to discuss mental activity in its public or communicative form, in its neurophysiological forms and, as well as a vocabulary to describe inanimate objects in a spatial concept.

I searched, in particular, for a theory of intelligence that explores all kinds of intellectual performance. I turned to two sources, *The Nature of Human Intelligence,* by J. Paul Guilford (1967), and *The Structure of Intellect: Its Interpretation and Its Uses, (SOI)* by Mary Meeker (1969).[2] Guilford proposes a three-dimensional model of intelligence that includes more than 120 factors rather than a model that is based on a single factor or a single dimension. Meeker, on the other hand, works her model into a form that is useful to teachers by constructing a three-dimensional matrix that describes a way of representing mental activity in its public form.

I used Meeker's SOI model rather than other competing models of intelligence because the SOI model describes intelligence in terms of mental activity (cognition, memory, divergent production, convergent production, evaluation), content (figures, words, symbols, and behavior), and product (units, classes, systems, relations, transformations, im-

[2] The Structure of Intellect is a theory of human intelligence that was developed from 1945 to 1965 by Dr. J. P. Guilford and his colleagues at the University of Southern California. The SOI is the application of the Structure of Intellect theory to various learning situations. It has been in continuous use and development from 1965 to the present.

plications) and such a construct was consistent with my mental picture.

Guilford's model—an attempt first to classify a spectrum of public disclosures—the "stuff" we perceive externally to ourselves—augmented the Meeker concept. The SOI model gave me a method for describing the public forms of mental activity. For example, in terms of Guilford/Meeker's model, the black and white chips in my dream are figural units. SOI notation then allowed me to describe how the teacher could operate on these chips in a manner that expresses some aspect of mathematics as he or she understands it. By using SOI notation I can express the public form of understanding the chips as cognition and memory of a figural unit. Similarly, in SOI notation when the teacher attaches understanding of, let's say, the property of additive inverse to the chips, it is a divergent production of a relationship.

Thus, I propose that objective space is occupied with physical or tangible objects and that these tangible objects filling the spaces can be described in SOI notation. Posters, pencils, and the set of black and white chips then share a common means of expression. The line of reasoning can be extended to include all objects that are perceived within the boundaries of the classroom, so that we now have the means to describe the classroom's physical or objective environment. I call the space in which the objects are defined or characterized in the above manner the Guilfordian space.

The Fabrication of the Guilfordian Space

The Guilfordian space contains objects from three classes. Some of these objects belong to a class that has permanency, e.g. chairs. Such objects are tangible and generally visually distinguishable. A second class of objects tends to be temporary or tends to dissipate, e.g., sounds. Finally, there are objects that form a class that is evocative, e.g., non-verbal communication.

These classes of objects—tangible, temporary, and evocative—collectively form the fabric of the Guilfordian space. An enriched environment is populated with many objects and can be very dense. There has been extensive research on the effects of enriched and deprived environments on learning in general and with experimental rats in particularly. Other studies have shown the effects of enriched environments on brain growth in small animals.

The "Bounded-ness" of the Contents of Mental Activity

For me, the notion of "Guilfordian" space provides a viable context to describe the external objects in space, though it does not describe how the brain deals with "mental activity." Mental activity includes: how the brain receives or constructs images of concrete objects (e.g., the physical chips); an internal symbol system (e.g., the teacher's knowledge of mathematics); and internal symbol systems associated with others to convey or show like-ness (e.g., when our teacher uses black and white chips to illustrate various properties of the set of integers). I consider each of these conditions as examples of the contents of our mental activity.

Mental activity is associated with the brain's capacity to integrate multiple sensory inputs (what we see, hear, smell, and feel) and, what we believe to "know" (has meaning) by complex processes that allows the brain to "host" and provide a permanent "residence" for these inputs (as content). I also think that the brain spends a great deal of time recognizing and reacting to the contents of our mental activity as new sensory inputs combine with them. When we recognize and react to the contents of our mental activity, I believe we are referring to the widely debated and undefined concept of "consciousness."

Although I have a rough idea of what is meant by "being conscious," consciousness is not something that occupies my thinking or concerns about my students. I assume that because my students are in my class that they are conscious. On the other hand, understanding how attention influences learning and how to compete for it is very important to me. I also assume that attention is a subset of consciousness. The expression, "please give me your undivided attention" suggests that students are aware of their surroundings and can choose to attend selectively to the contents of their mental activity and the external environment. I presume that when students are aware of the content of their mental activity it is verbally reportable—for example, when one student explains to the class how he or she solved a math problem that was completed for homework the night before. A student is considered conscious of the contents of his mental activity when 1.) he or she shared the content at a particular time, indicating that his or her sensory receptors are tapped into a shared time frame, and 2.) that he or she was able to report verbally about thoughts that occurred during the shared time frame.

I also think consciousness, in the context of teaching, is when students are aware of some information and react to something externally on the basis of that information. For instance, when a student observes that a computational error was made during an illustration of a particular algorithm demonstrated by another individual, indicates that he is cognizant of his own information and participated in the interaction with at least one other individual.

Attention is more observable than consciousness in classroom situations because teachers generally treat levels of on- and off-task behavior as an indicator attention.

Boundedness of Mental Activity

I hypothesize that the contents of mental activities are "bounded." I use the term "bounded-ness" to mean an implied differentiation. The results of my research on consciousness suggest that my notion of bounded-ness, or an implied differentiation is substantiated. There are two competing approaches to the study of "consciousness" from the field of neuroscience where I found support for my notion of bounded-ness of the contents of mental activity.

Single Neuronal Correlate of Consciousness

One approach (Crick and Koch, 1995) asserts that we are not directly aware of our inner world of thoughts, intentions and planning (that is, of our unconscious homunculus[3]) but we are aware of the sensory representations associated with these mental activities (my content of our mental activity). Those in support of this experimental paradigm actively pursue the idea that a mechanism for a simple form of consciousness, visual awareness, might be found with just a few brain regions (at the neuronal level), the closely linked thalamic[4] and visual cortex regions (Crick and Koch 1998).

The source of data useful for this type of analysis of the "neuronal correlate of consciousness" generally comes from experiments with 1) cases of human brain damage that alter

[3] The homunculus is usually thought of as a "little man inside the head," who perceives the world through the senses, thinks, and plans and executes voluntary actions.

[4] The thalamus is a relatively large collection of cell body clusters shaped like two small footballs. It is involved in receiving sensory information from the eyes and other sense organs, processing that information, and then transmitting it to primary sensory zones in the cerebral cortex. The thalamus also processes pain signals from the spinal cord as well as information from different parts of the cerebral hemispheres, and relays it to the cerebellum and the medulla. Together with the hypothalamus, the thalamus forms part of the forebrain called the diencephalon.

By registering the sensory properties of food, such as texture and temperature, the thalamus plays a role in appetite. It is also known to be involved in the control of sleep and wakefulness. Cognitive researchers have found that the thalamus activates or integrates language functions, plays a role in memory, and that a portion of the thalamus, called the pulvinar, helps in refocusing attention. Together with the hippocampus and parts of the cortex, it is instrumental in the formation of new memories, which are then thought to be stored in the cerebral cortex.

consciousness, or results from lesioning studies in animals, 2) recordings from animals which are used in attempts to identify specific types of neurons with activity patterns that match the perceptual states of the animals, and 3) electrical stimulation of specific brain regions that seem to be able to shift an animal's conscious perception of a stimulus.

I understand the underlying assumption of these experimental approaches is that visual stimuli provide a rich array of inputs to the nervous system, such that they register with greater frequency and strength than when the visual stimulus is not the content under consideration. When specific content of the visual stimuli produces different firing rates in the brain from content that is not in the emphasized field of vision, it is believed that these results are examples of neuronal correlate of consciousness, or as the NCC. Hence, NCC is when neural firing directly correlates with the perception of the object that one is focusing on at the time (Koch and Crick 1999).

From a theoretical perspective, the central question of those in support of NCC is that a neuron can "learn" the visual representation for natural stimuli through experience. I take the meaning of learning in this case as referring to differences in the levels of frequency of neurons' (and associated cells) firing rates that are produced when the subject is looking at a particular object than when he is not.

The experimental procedures employed to study the above question is a procedure called binocular rivalry. In binocular rivalry, a subject views two incongruent visual stimuli, one through each eye, simultaneously, but consciously perceives only one of the stimuli at a time. Every few seconds, there is a switch in perceptual dominance, and the subject perceives the other stimulus. Neuroscientists have found that in spite of the constant visual stimulus, observers see the horizontal grating alternate every few seconds with the vertical one suggesting that the brain does not allow for the simultaneous perception of both images (Blake 1989).

Related to the horizontal/vertical grating experiment, Sheinberg and Logothetis (1997) recorded neural activity in the inferior temporal cortex where subjects had a flashing image in one eye while the second eye was fixating on another pattern, and found that approximately 90% of the recorded neurons were found to reliably predict the perceptual state of the animal. Moreover, many of these neurons responded in an almost all-or-none fashion, firing strongly for one percept, yet only at minimal level for the alternative one.

In both cases it was found that for very small increments of time the brain had to relinquish the "seeing" of one percept in order to attend to the other. For me these findings represented the first indication that my hypothesis concerning the "bounded-ness" of the contents' mental activity, or an implied differentiation, may be a valid and useful construct.

Binocular rivalry was also used in a series of experiments with macaque monkeys by Logothetis and colleagues (Logothetis 1998), as they recorded from a variety of visual cortical areas in the awake state whether the monkey observed the left or the right image while performing a binocular rivalry task. In their analysis of these data they found that in a high-level cortical area such as the inferior temporal (IT) cortex, almost all neurons responded only to the perceptual dominant stimulus. In other words, a "face" cell only fired when the animal indicated by its performance that it saw the face and not the pattern presented to the other eye. These findings further support my hypothesis for an implied differentiation, or "bounded-ness" of the contents of mental activity because certain neurons respond to the perceptual dominant stimulus but not other stimuli.

I also found evidence in support of my concept of implied differentiation, or "bounded-ness" of the contents of mental activity in the Crick and Koch (1998) hypothesis of the single neural correlate of consciousness when they say that "the neuronal basis of consciousness is the notion that minimal neural activity is sufficient to cause a specific conscious percept or memory." This implies some type of differentiation, or bounded-ness.

Language processing has also been studied by neuroscientists. Employing what is referred to as Single-Photon Emission Tomography (SPECT), researchers are trying to demonstrate that language processing is localized in areas of the brain by studying blood flow there. These studies are typically performed with the patient in a resting state so as to minimize confounding effects of activity.

Applying imaging techniques, Lassen found that the cerebral cortex generates recognition patterns in the form of concentration of protein in a specific area of the brain when an individual hears his own voice and does not generate recognition patterns when he hears a language he does not understand—a differentiation. Therefore, the sound of the words "la casa" would not elicit a pattern in the individual who does not understand the Spanish language. But, if one showed that person a picture of a house and then spoke the sound associated with the letters "la casa," then one should see the recognition patterns. Moreover, these recognition patterns should then be unique to the sound associated with the letters "la casa." This research establishes, for me, the notion that mental activity associated with auditory brain functions also shows signs of a differentiation—a "boundedness."

Blake (1989) and Lozanovs also found that the brain does not allow for the simultaneous perception of two images. In particular, Lozanovs found that it is virtually impossible to attend to two conversations simultaneously, but that the listener can alternate between the two, storing information along the way. The fact that a listener must alternate between the two conversations implies, again, that one must be relinquished for the other as the listener shifts attention back and forth between the two.

The research from NCC provides more preliminary evidence in support of my notions that the contents of mental activity are bounded. Further evidence in support of my notion is provided at the neuronal level by Crick and Koch (1995), postulate that to attend to an object the brain has to construct a multi-level, explicit symbolic interpretation of part of the visual scene.

Others in the field neuroscience believe that it is not possible to understand consciousness by assuming that there are single clusters of neurons that directly correlate with what we see. Instead, the problem of consciousness should be studied from a more global perspective. In other words, consciousness involves many groups of neurons. Some happen to be configured in a way that makes them respond more strongly than the others to a specific stimulus. Neurons that respond to the stimulus better are selected by the brain because for some unknown reason they are best qualified and react to the stimulus more strongly. As this group of neurons is strengthened with the particular visual stimulus, they will eventually represent the particular visual stimulus—the content of the mental activity.

Theory of Neuronal Group Selection

One alternative approach for the study of consciousness is the theory developed by Gerald Edelman (1987, 1992) and used by scientists like Gazzaniga (1985); and Paivio (1986). The Theory of Neuronal Group Selection (also called Neural Darwinism) was developed by Edelman in an attempt to explain consciousness from a biological perspective. This theory proposes that consciousness is a global brain function, meaning that many widely distributed parts of the brain are important for consciousness and that no single group of neurons represents what we see.

The explanation of the Theory of Neuronal Group Selection is described in a book by Gerald M. Edelman entitled, *Bright Air, Brilliant Fire: On the Matter of the Mind.* Edelman contends that the world does not come pre-defined to us, but that it is through our sense organs, as we move about taking samplings of the world and creating what he calls categories, or maps. These are constructed soon after birth and are used to construct the foundations of our sense of reality and to establish foundations of meaning of the contents of our mental activity.

The Theory of Neuronal Group Selection describes a brain that first begins to build a structure in which complex functions of the mind are used to construct our reality of the world. The smallest element of this structure is a neural group of cells which are called a category, or a map. A "map" is an interconnected series of neuronal groups that respond selectively to things which are found in the visual world (movement and color) and represent our experience of them and are internally linked together.

Edelman believes that: "these categories are not individual nerve cells or neurons, but groups ranging from about fifty to ten thousand neurons. During the development of the fetus, a unique neuronal pattern of connections is created, and then, starting with the infant stage, experience acts upon this pattern, modifying it by selectively strengthening or weakening connections between neuronal groups, or creating entirely new connections."

I interpret this statement as suggesting that the first level of differentiation of contents of mental activity is the formation of sets of neurons that are based on the child's experience of seeing, hearing, and touching the "stuff" from the external environment, and that they (neural groups of cells) are linked because things like movement, color, proximity, separation, order, and enclosure represent the basic elements of our mental activities. Which groups of neurons that responded to the stimulus is not pre-determined. Random groups of neurons respond because it is believed that they are the fittest for this particular stimulus. Because these groups respond to the particular stimulus that have somewhat of an edge on others when the stimulus is revisited by the individual, it then becomes the unique host of this particular stimulus.

This aspect of the theory, I believe, is consistent with Piaget's (Piaget, 1967) thoughts on how children in the perceptual or sensori-motor stage develop their notions of what he calls perceptual space, or representational space (p. 3). Piaget, like Edelman, believes that our initial notions of space do not represent a mere copy of existing sensori-motor constructs. He writes,

> This sensori-motor space is superimposed upon various pre-existing spaces, though it is by no means a simple reflection or repetition of them. On the contrary, it has its own course of development which can be traced fairly easily, and in addition, the spatial organization or sensori-motor behavior results in new mental constructs, complete with their own laws. (p 3).

Edelman believes that the interrelation among the initial categories (maps) suggests that there may be some overlap between series of neuronal groups, which further suggests that these categories participate in several firing sessions in response to the visual world. In other words, these groups of neurons are preparing the brain for "structure" by first forming these rudimentary perceptual categories, but also that these groups may participate in other instances of random neural group selection situations. Because objects are not "labeled" (they do not come "already divided into objects") we must make them, in effect, through our own categorizations—"Every perception is an act of creation." Object constancy is perhaps achieved due to the strengthening of interconnections of neuronal groups which form mental categories for our sense of reality.

Piaget contends that "the perception of space involves a gradual construction, and certainty does not exist ready made at the outset of mental development" (p. 6). He then suggests that this construction consists of *proximity,* the "nearby-ness" of elements; *separation,* the means by which objects are dissociated; *order,* how elements in the visual field are ranged one before another; *closure,* where a percept is enclosed along one dimension, surrounded by another in two dimension, or inside as in the case of an object in a box are the rudimentary elements of mental activities that are developed, or formed during the early stages of childhood (p. 8).

Edelman also believes that every perception is an act of creation and that there are no specific sets of neurons linked to our perception (i.e., a baby's bottle). I think that Edelman would argue that object permanency is a function of several groupings of neurons, connected and working together to help form our understanding of "bottle-ism." And that with greater frequency of "seeing" and "handling" the feeding bottle, permanency, in a neural sense is then achieved through a process referred to as "re-entrant signaling." Re-entrant signaling is the process that allows a new experience to enter older categories (maps) and modifies them—a continuous "communication" between the active maps themselves, which then enables a coherent construct such as "bottle-ism" to be made.

Our construction of "bottle-ism" arises from the interaction of many sets of neural groups. That is, stimuli from, say, touching a bottle may affect one category and stimuli from seeing it may affect another set. Consequently, re-entrant signaling takes place between the two categories as part of the process of perceiving a feeding bottle.

Piagetian theory suggests that "the constancy of shapes results from the sensor-motor construction at the time of the co-ordination of perspectives. During the sensori-motor stage objects change their perspective and such alterations are perceived not as changes in the point of view of the subject relative to objects, but as actual transformations of the objects themselves. His example in support of his interpretations of a young child's behavior is:

> The baby waggling his head before a hanging object behaves as if he acted upon it by jerking about, and it is not until the age of about 8-9 months that he really explores the perspective effects of actual displacements. Now it is just about this age that he is first able, for instance, to

> reverse a feeding bottle presented to him wrong way round. That is, to attribute a fixed shape to a permanent solid. (p. 11)

The Theory of Neuronal Group Selection (Edelman) suggests that a "bounded-ness" of mental activity may be hypothesized due to the manner in which the brain forms what are called perceptual categories (maps) of primary consciousness. According to the theory, groups of neurons categories (maps) are interconnected in the brain from various sources (movement, color, touching, seeing) to construct our sense of reality (e.g., our notion of chair-ness). Piaget's theory on how children develop their conceptualization of perceptual space is compatible with Edelman's theory of consciousness.

Where in the Theory of Neuronal Group Selection do the notions of "bounded-ness" resides? For me, Edelman's use of the term category (a map) implies a "bounded-ness" when he says the following:

> the concept of "a chair" is constructed over time through a series of strengthening selected neuronal maps and scene and then refined through a process of re-entrant signaling, there is an alliance of neurons constantly forming and shifting boundaries, providing an ever-changing population of neuronal groups available for selection. They are either a category or represent a categorization.

To my satisfaction both approaches to the problems of consciousness have suggested that there is reason to believe that the brain does show signs of a differentiation between the contents of mental activity. The evidence from NCC is more direct—visual perception was shown to be directly correlated to single groups of neurons in the brain. The "bounded-ness" of mental activity is implied because it was found through experimentation that one set of such neurons will be all of one type, will probably be fairly close together, and will all project to roughly the same place (Adolph et al, 1994).

I chose to learn about Edelman's Theory of Neuronal Group Selection because it represents a major contribution to the field of neuroscience and it is an alternative to the approach that is used by supporters of NCC to study consciousness. It was also important for me to determine, as I suggest, whether my notion of bounded-ness of the contents of mental activity is valid. The features of the Theory of Neuronal Group Selection that suggest that there is a bounded-ness of the contents of mental activity are, first, there is formation of fundamental categories (maps) of primary consciousness that develops very early in a young child. From the sampling and re-sampling of the environment these fundamental categories are modified and strengthened to form more complex categories.

A category, according to *The American Heritage® Dictionary of the English Language*: Fourth Edition 2000 is a general class of ideas, terms, or things that mark divisions or coordinations within a conceptual scheme. For example, Aristotle uses *category* to express various modes of objective being, such as quality, quantity, or relation that are inherent in everything. Then we have Kant's use of the word as to suggest various modes of subjective understanding, such as singularity, universality, or particularity that organize perceptions into knowledge.

It is from Edelman's descriptions of how categories (maps) are formed in the brain and from the definitions the word provided by the *The American Heritage Dictionary* that I feel comfortable with the notion that there is differentiation of the contents of mental activity.

In pyschological research, the "bounded-ness" of the contents of mental activity has formed the basis for landmark findings in several areas of thought and investigation. For example, J. G. March and the Nobel-prize winning H. A. Simon (1958) studied organizational and economic decision-making and concluded that people are "boundedly rational." In other words, individuals make decisions with available information, so that a person is only operating within the boundaries of objectivity with the information on hand at the time. Lacking total information one is placed in a position of being partly objective and partly subjective. Bounded rationality, to me, is an implied differentiation of the contents of mental activity.

Another example of "bounded-ness" is found in the literature about how teachers plan and conduct instruction. The rational model of teacher planning first proposed by Ralph Tyler (1950, 1970) recommends four fundamental steps for effective planning: (1) specify objectives; (2) select learning activities; (3) organize learning activities; and (4) specify evaluation procedures. This model can be characterized as a rational means-ends model in which the first step is to decide upon the ends (objectives) to be accomplished, and the succeeding steps involve the choice of means (activities) and assessment procedures for measuring the attainment of the goals.

The rational planning model is challenged with evidence of how teachers report their actual planning strategies. Zahorik (1970, 1975), Yinger (1978), Clark and Yinger (1979) and Peterson and Clark (1978) found that learning objectives are seldom the starting point for planning. Rather, teachers seem to plan around the activity. Taylor (1970) and McCutcheon (1979) found similar results in their studies. The activity of planning is a "bounded-ness", in the mind of the teacher, is the essence of an implied differentiation. The activity is also bounded because it is the activity that prescribes those behaviors (cognitive and affective) that will yield the expected outcome. The teacher is likely to steer the students back to the original plan should they deviate, the implication being that there are responses that are inappropriate for the particular lesson, at least according to the teacher. In sum, then, evidence suggests that the concept of "bounded-ness" operates on several levels of mental activity and has been noted and remarked upon in various areas, from work of neuro-physiology to political decision-making—from the micro-level to the macro-level.

To return for a moment to the dream situation: both teacher and student interact with the environment and respond to their own internal content in constructing the individual reality of the environment.

The student perceives the set of black and white chips in the room. Since every perception is an act of creation, according to the Theory of Neuronal Group Selections if, and hopefully when, the student explores the chips he is seeking (or imposing) meaning in them. Through these interactions he adds to his personal reality—the sum total of his mental activity.

The teacher constructs his sense of reality in the same way. For a moment let's focus on the subset of the teacher's mental activity that specifically includes: the content of his images of the physical chips, the content he views as his knowledge of formal mathematics, and the content of his analogy between the black and white chips and his knowledge of the set of integers—and, that the teacher has decided to share the analogy with the student.

I propose that there is no way for the teacher to communicate his or her full understanding of, for example, the property of additive inverse with the student in a single manifestation—sensory outputs. Instead, he will reveal the contents (his understanding) to the public, by "puncturing space" with segments of the particular idea. For example, the teacher intends to communicate his, or her understanding of the property of additive inverse (mental activity) by puncturing space (an insertion) with segments of this understanding through both written and spoken language, diagrams and visual models, and in some cases from his or her facial and body language. Collectively, these things represent his full understanding of the concept.

Consequently, the students receive parts of the information as manifestations of the teacher's content knowledge. For the student to accept these parts he must first relinquish her or his own mental activity in order to shift his attention to the incoming stimuli–bounded-ness, or implied differentiation of his mental activity. This is why I feel, and say, that to teach is to compete for the students' attention with the internal mental activities and the external stimuli from the environment.

If the teacher chooses to communicate understanding of additive inverses through interactions with the set of black and white chips, the objective is to help the student find meaning in them and hopes that the student will begin to perceive the set of chips in a similar way. Teaching strategies establish meaning in the chips and produce a convergence towards perceptual and conceptual identity because the meaning that is sought is the meaning that the teacher accepts. Verification that convergence towards perceptual and conceptual identity is in process is achieved through various external indicators—tests and other forms of assessment.

We have seen that the notion of a Guilfordian space has provided a conceptual framework within which to examine the interaction between student and teacher. In addition, the SOI model, in an adapted form, provided us with a way of conceptualizing mental activities and the space that houses them. "Puncturing space" is a strategy that is used to initiate the convergence toward perceptual and conceptual identity between two, or more individuals. Space is punctured with objects. Most of the time space is punctured with combinations of objects, physical (a chair), words (concepts), and kinesthetically (body language).

The Lesson and the Lesson Plan as a Pedagogical Object

A special type of object found within the Guilfordian space is that which forms what could be called the pedagogical domain. As the phrase suggests, the objects of this sort are designed or constructed to stimulate mental activity and to foster objects' movement between subsets of the overall mental activity structure in the context of the classroom and school settings. An excellent definition of the pedagogical domain is offered by Hudgins (Bryce Hudgins, 1974) who defined the term as:

> those settings and those people that influence the content which is to be taught to pupils, and those methods or procedures employed for purposes of conveying content. The domain includes decisions about the choice of content and its communication.

Following on from this, one can posit that the fundamental tool of pedagogy is the pedagogical object. I propose that when teaching occurs, the pedagogical object is used to puncture the Guilfordian space. The lesson plan is the first element, of the pedagogical object, a construct, a statement of purpose and an outline of prompts. It contains the part of the pedagogical object that can be described in written form, but it does not contain the affective elements of the lesson itself—tone of voice, appearance, facial expressions, or the emotional state of the teacher. The actual lesson—the manifestation of the lesson plan and the second element of the pedagogical object—is, then, the expressed form of the purpose.

As observers of the hypothetical interaction, we can assume that the purpose of the lesson and the prompts that guide students to achieve the objectives, are parts of the teacher's mental activity. According to this theory, the lesson, or the second element of the pedagogical object, as the expressed form of the purpose, is manifested in segments. As we watch the teacher puncture space with segments of the pedagogical object, we can sense whether or not the sequence of prompts helps the student approach a perceptual and conceptual understanding similar to the teacher's. Clearly, whether the student can, in fact, follow the sequence with understanding is, at least in part, due to what he, "brings to the table." But, in any event, it is the order and manner in which the teacher manifests the parts of the mental activity that should allow for the greatest number of individuals to understand the lesson, that is to achieve a convergence towards perceptual and conceptual identity with the teacher. The same can be said for this argument itself, if this paper is written in a fashion that presents the thesis in a logical way, then at the end the reader and the author will have converged toward a conceptual identity. Whether the reader agrees with the author or not is another issue, but there should be enough understanding conveyed to enable the reader to form an opinion.

The conceptual framework of the Guilfordian space enables one to view both psychological and physical space in terms that attach new descriptions to what we view as teaching by providing a vocabulary of concepts and a structure in which to study these concepts. Teaching can then described as puncturing space with pedagogical objects.

Pedagogical objects are the expressions of purposes and are bounded in the sense that multiple expressions cannot exist (be processed) simultaneously by a single individual. Empirical evidence suggests that the objects tend to strengthen, or weaken the internal mental activity of the recipient. Thus, pedagogical objects as external stimuli compete with other external stimuli, and with the recipient's internal mental activity, for the "attention" of the student.

As a result of my extensive period of thinking, a set of ideas has become fundamental to my teaching:

- Human mental activity is occurring at all times.
- Attending to mental activity external to one's self forces the bounding of one's own internal mental activity.
- The public forms of all mental activities are combinations of operations, contents, and products that punctured Guilfordian space and when inserted into the Guilfordian space, the public form of a mental activity is bounded, by exclusion. That is, the mental activity is made public for a purpose. Purpose is exclusionary, and is bounded—an implied differentiation.
- The Pedagogical Domain comprises the settings and people that influence the content that is to be taught. The pedagogical domain includes decisions about the choice of content and its communication.

- Pedagogical objects are the methods or procedures employed for purposes of conveying content and stimulating thinking.
- Teaching is the puncturing of space with pedagogical objects that compete for the student's attention with the internal mental activity of students and with other external stimuli.

Chapter 4

Resolution

It is all well and good to create a hypothetical conceptual framework, but it then becomes critical to ask how that framework—"puncturing space with pedagogical objects"—directly influenced my practice of teaching mathematics? In fact, it has done so in several ways. The notion of puncturing space forces greater self-awareness of one's total "being" expressed through one's facial expressions, body language, words spoken, and physical objects. In short, it influences the entire set of cognitive and social/linguistic interactions that occur within the classroom environment. When teachers walk through the door of their classrooms, they first create an "attitude" in the classroom (we generally call this setting the tone). For example, if the teacher is carrying a pile of papers, then students conclude that the teacher is returning a set of exams or a written assignment, or will distribute a new assignment or examination,[1] and their mental state of anticipation adjusts accordingly.

Employing the conceptual framework of puncturing space also changes the teacher's perceptions of the learner and, in turn, those perceptions condition the approach to teaching. I recall that during my first three years, when much

[1] First year teachers struggle with this concept of "setting tone." I feel that if student teachers and first year teachers think about this concept in terms of puncturing space they will develop a better understanding of their power to manage their classrooms.

of my effort was devoted to understanding the culture of the school, I used a textbook and lectured. My responses to students gave them few opportunities to think for themselves. Nothing about my classroom practices allowed students to find meaning in their own work. Consequently, students learned that they would be rewarded if they sat and listened passively to my understanding of mathematics. But, when "puncturing space with pedagogical objects" became the framework in which I thought about and practiced, my teaching, and when "unlocking potential" became my goal, the dynamics of classroom interactions changed significantly.

But, just what is meant by unlocking students' mathematical potential and when do we know that we have succeeded? When I say we seek to "unlock students' mathematical potential," I mean that students came to the classroom with a "store" of knowledge—their potential—yet, are often unaware how to employ that knowledge to learn new mathematics. The act of enabling students to recognize the knowledge they possess and to relate that knowledge to new information is what I mean by unlocking their potential. A curriculum that provides situations in which students experience this recognition and internalize these relationships is one that unlocks students' intellectual potential. If students explain their answers and acknowledge how they constructed mathematical arguments, their teachers have some evidence that students have unlocked their potential.

For example, suppose that sixth graders are asked to combine the algebraic expression $1/a + 1/b$, they already know, from arithmetic, how to add positive rational numbers with unlike denominators. The first step in understanding the algebraic problem is to create scenarios in which students recognize that adding 1/a + 1/b involves the addition of positive rational numbers. Students' potential is unlocked, when they recognize that they can apply some of their knowledge about adding fractions with unlike denominators to combining the algebraic expression, and then determine what is needed to bring closure to this problem.

Structuring math activities around questions and broad concepts renders teaching more constructivist in that it directs students to survey their knowledge base to locate "chunks" of information that apply to the new problems. Students figure out what they already know or understand about a problem, use this information to begin problem solving, and then speculate about what is needed to further their progress. Often students will not immediately recognize that they possess relevant information, but when students fail to identify useful information, the use of brainstorming sessions to help them locate this often shows them that they know more than they had assumed. The teacher's role in the brainstorming sessions is to pose questions that initiate general discussion about the next steps in the thinking and problem-solving process. When necessary, we introduce new vocabulary and help students attach meaning to the new words. For instance, I have often asked during class discussions, *"Is there something in the work we did prior to this problem that will help us to determine what we should do next or think about?"* Fully realized constructivist teaching trains students to initiate such prompts for themselves. When they are permitted to learn at their own rate and to view mistakes as learning opportunities, students are put in the role of “makers of meaning.”

The power of the conceptual framework, puncturing space to unlock the potential of students, can be illustrated by the following story about a co-teaching experience that I had years ago at the Dalton School in New York.

In September of 1990 I taught math to sixth graders in a team teaching arrangement with a colleague, K. Allen. K. Allen had a remarkable talent for analyzing complex algebraic problems numerically. I, on the other hand, am a visual learner and thinker with an aptitude for analyzing complex algebraic problems geometrically. She considers herself a numerical thinker, and I consider myself a geometer. The differences in problem solving style are clear in the following example of a problem that we used in a math assignment for our students.

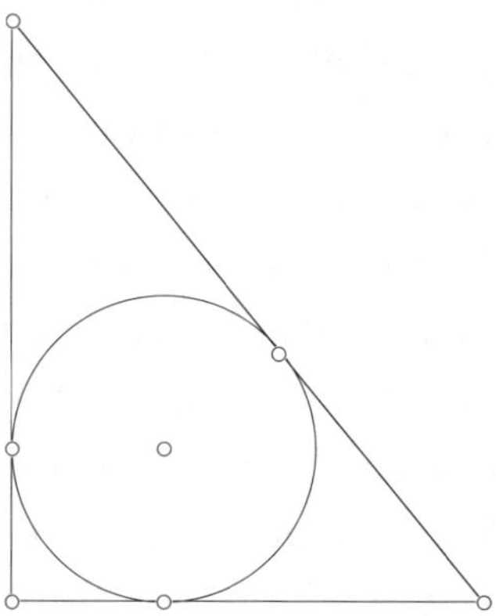

A circle is inscribed in a right triangle. Show that the diameter of the circle is the difference between the sum of the legs and the hypotenuse.

Of course, before we gave students problems to solve, we solved them for ourselves. K.Allen structured the problem within her intuitive understanding of number patterns that are derived when they are dependent on a mathematical relationship or function. She then solved the problem by assigning numerical values to each segment of the right triangle, made a table of her computations, and then generalized her findings. I relied on my extensive knowledge of geometry to structure this problem and deductively derived the stated relationship algebraically. She tended to generate sets of numbers to allow her to look for patterns; I drew diagrams.[2] Although K. Allen and I solved geometry problems from different perspectives, our common ground was found in our appreciation and understanding of mathematics, respect for differences in learning styles, and a firm belief that there is much to gain from theories of learning and teaching.

During our planning sessions, we agreed that the "question" would be the leading pedagogical object used to puncture the social-linguistic space of our classroom. We wanted students to pose questions, when they were able to, and we agreed that we would intervene only when students had reached mental blocks in their thinking. We felt that we could determine when students had reached a mental block by noting when they began to act "silly." The social-linguistic interventions that are described above is what Vygotsky would characterize as teaching within the "zones of

[2] Because our approaches to problem solving were different, our students had a choice of whom to seek out for help. Gender did not appear to be a critical piece in their decision.

proximal developments" of students. A zone of proximal development is defined as the difference between what a child could accomplish unassisted in problem solving and what he or she could accomplish only with assistance. Vygotsky argues that socially-supported activity in the zone of proximal development awakens and provides paths for intellectual development.

One of our three math classes became interested in determining whether mathematics was "constructed" or "discovered?" The question was framed by posing a hypothetical situation involving two shepherds at a time when formal language and symbolic representations of numbers had not been developed formally. The two individuals needed to determine who had the greater number of cattle. We punctured space with a "mind experiment." A mind experiment is bounded by the parameters of the situation—a lack of formal language and symbolic representation of numbers. Our purpose was to get students to conceive of how the two shepherds could determine which one had the greater number of cattle. The students had to determine what they understood about the problem at the outset, use that information to begin problem solving, and finally speculate about what was needed to further their progress. We also wanted students to explain their answers and articulate how they constructed mathematical arguments about their answers.

As we listened to the students' initial discussions, it became clear that they had forgotten much of the pre-operational mathematics that they had experienced while in the lower elementary school, and in particluar, the concept of equivalent sets. To remedy this, K. Allen and I reiterated these early mathematical experiences in a manner that sought to attach them to the abstract mind experiment by inserting the problem into a concrete host.

Reduced to its bare minimum, the shepherds problem falls into the area of math known as set theory. In conceptualizing the problem, sets of cattle can be treated as sets of chips. The sets of chips and the rules governing their combination serve as a conceptual proxy of the mathematics that is implied in the mind experiment. That is, what is learned in the manipulation of the chips represents what could be learned by manipulating the elements of the abstract mind experiment. This is an illustration of Piaget's concept that logical thinking can be attached to concrete manipulations.

In our class, we employed the chips to review elementary set theory, and particularly the concepts of equivalent sets, one-to-one correspondence, and the notion of "less than," and "greater than."

The manipulations of the chips allowed students to discover the concept of "next after numbers." The "next after number" concept developed out of discussions led by K. Allen and me about a story that involved the set of black and white chips. Each student was asked to create a set of black chips that represented their favorite natural number, but to hide the set from the rest of the class. Some students covered the sets with a piece of paper; others hid them underneath the table. A series of questions (a form of pedagogical objects) was then asked. For example:

QUESTIONS: Can someone determine one thing that is common to each set of chips no matter how large or small it is?

Silence infused the class for several minutes. K. Allen and I decided to leave the room so that the students could discuss the question in their own psychological space without our interference. When we returned Jenny F. raised her hand and replied, "Each of the sets has a 'before it' set of chips and an 'after it' set of chips."

QUESTION: What do you mean?

JENNY: Four comes before five, and six is after five.

QUESTION: If we ran up and down the number line does your "understanding" of this idea always work?

JENNY: Yes.

QUESTIONS: Can you frame your "understanding" of this idea in a way that we can use to test whether each of the sets of chips has a before and an after set?

Once again there was silence in the class, but this time it was very different. This silence suggested that the students were stumped. Apparently they needed help interpreting the question. K. Allen and I had entered their zone of proximal development according to Vygotksyan theory. Recall that the zone of proximal development is defined as the difference between what a child could accomplish unassisted in problem solving and what he or she could accomplish with assistance. To advance the discussion, we asked a follow-up question, hoping that the question would break the block and help the students reestablish their thinking.

QUESTION: What part of your favorite set of chips is in the next after set?

JENNY: All of it?

QUESTION: All of it, and what?

JENNY: All of it plus one.

QUESTION: How much of your favorite number is in the before set of chips?

JENNY: All of it. But, the before number is my favorite number minus one.

QUESTION: Does anyone know how mathematicians would express or represent an unknown idea, or number?

JENNY: With a letter?

QUESTIONS: We call these unknowns. What letter or symbol would you like to use to represent our unknown favorite chip set?

JENNY: x.

QUESTION: If x represents the chip set for each of us, then how do we represent our next after chip sets.

OTHER STUDENTS: x+1.

QUESTIONS: If we know x + 1, can we use that expression to find someone's favorite chip sets? How?

OTHER STUDENTS: Yes, by subtracting one from it—(x+1)-1.

Furthermore, because x could take on any value, the group concluded that (x) and (x +1) collectively symbolized the mathematics of chasing of infinity. Where (x) was the frame which could hold an unknown number, and (x +1) was the expression in which the largest possible number one could think of is pushed to its "next after" number.

The result of this discussion was that when we returned to the mind experiment of the two ancient shepherds, the students were able to think about this problem logically because it was now attached to a concrete example—a proxy. K. Allen and I often asked our students to write math essays on various topics, problems, or experiences that they had and Jenny F. wrote an essay to describe her approach to the shepherds' problem that demonsatrated her ability to link the network of concepts that developed during our class discussions. It represents her ability to link a network of concepts that developed during our class discussions. This is an unedited version of the paper.

Is Mathematics Constructed or Discovered?
by Jenny F.

That is the question we set out to understand. However, one thing led to another, and many topics were presented throughout our discussion. There was a general agreement that mathematics owed much to both discovery and construction. Value has always existed, two has always had "two-ness," therefore man could not have constructed that aspect of mathematics. Sounds like, "one," "two" and "add," or names for values and operations, are not concepts that have always existed, and could not have been there for man to discover. Value brought us to greater and less, whom we analyzed through "The Cattle example."

We used one-to-one correspondence, or matching, to determine greater or less than in groups of cattle.

If set A has two cattle and set B has three cattle, assuming man has not yet named values (or had given them titles different than our current ones), How would either herds person prove that set A, had fewer cattle than set B?

We used the concept of one-to-one correspondence, or matching.

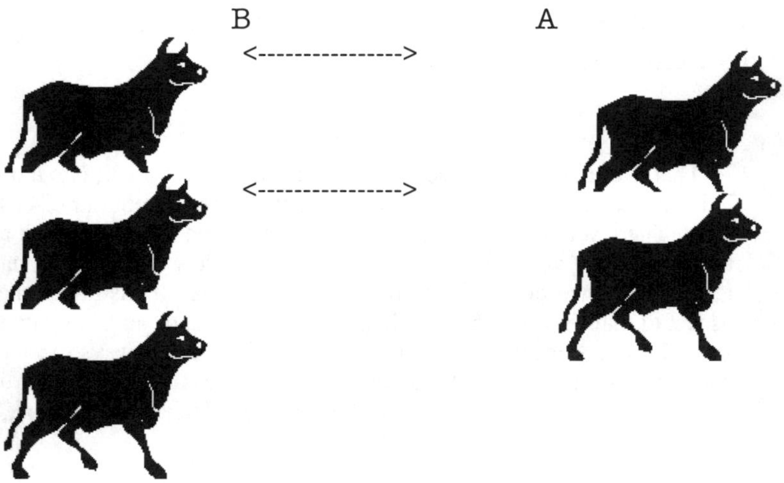

We saw that set A set B. We wondered what we could do to make A = B. We then thought that if Set A is less than Set B then there must be some number say "k" that could be added to Set A to make it equal to Set B. It would not make sense for k to equal zero.

$$A + k = B$$

"+1" made us think about "next after" (3 is next after 2, etc.). A mathematical way to say "next after" is x + 1. The math thing called x + 1 is closely related to the concept of infinity. We talked about infinity not being a particular place. We said that you can "chase" infinity by using x+1. The next question proposed was:

Does (1) A B, work for all cases where A B?

To answer this question we used x+1. We used x+1 because it is one of the best ways to represent chasing infinity on paper. If the highest number you can think of is x and you "+1" it, you have just pushed infinity a little further.

We applied x+1 to our statement

(2) (A+1) (B+1)

A+1 or B+1 is very similar to x + 1. We looked at how statement (1) is related to statement (2). This relationship showed us that no matter how large the numbers get, if you continue adding 1 to both sides, the greater set will stay greater. You will always have to add to the lesser set to make the two sets equal. So our statement proved to us that our A + k argument would work for all cases such that A was < B.

The approach to meaning-making taken by Jenny F. is typical of students who demonstrate a genuine interest in exploring mathematical questions. Most are intuitive thinkers. Talking with them, I often feel as though I am conversing with future philosophers. Their questions address abstract ideas and concepts, but it is worth noting that at this level of development they can pursue this level of thinking only when it is attached to concrete examples.

The questions posed by my students provide a clear indication of how relevance can emerge from students themselves. There were other lessons, as well—we observed that our students did not connect lessons on set theory that had been taught in primary school to the present discussion, although they did acknowledge the prior experience when it was pointed out to them. We learned that the intellectual practice of these particular children within the classroom community was the testing of their "spontaneous theories," spontaneous in the sense the students used the most current knowledge available to generate explanations. The classroom discussion and the spontaneous theorizing are examples of "bounded rationality" much as March and Simons have suggested.

We observed time and again how critical our choice of words and use of language was to the development of our students' thinking. And although our students selected the themes of the discussion, we found it necessary to help them maintain their focus and give them the language to maintain coherence in their thinking or intellectual practice. The writings on social constructivism by L.S. Vygotsky (1978) and his recent interpreters (e.g., Rogoff, 1990) argue that language is a primary vehicle for intellectual development. Language is not only a means of communicating information, but is also an important vehicle for helping learners broaden and deepen their understanding of important ideas.

This chapter began with the question: *how does the concept of puncturing space with pedagogical objects directly influence the practice of teaching mathematics?* The answer is that it empowers teachers with a framework for viewing the classroom environment and for managing behavior within this environment, and it allows a practical means of moving between abstract thinking and concrete thinking through the use of concept proxies.

Our success with the use of a concrete set of chips and the review of elementary set theory as a means to attach relevance to the "cattle" problem, and numerous other examples as well, increased my interest in the idea of pedagogical objects in general and concept proxies in particular. It is legitimate to ask how these ideas hold up when they are considered with traditional theories of learning.

It is believed by many that students at the concrete operational level of cognitive development can think logically, but only when they relate their mental activity to real objects in the physical world. Piagetian theory describes the concrete operational child's cognition as an "isomorphic," (i.e., having like characteristics) step-by-step mental replica of concrete actions and events. That is, the young child simply visualizes reality sequences in his or her mind, just as they might occur if the child were taking overt action. Comic "flipbooks" that children purchase in novelty stores provide a physical model that can help us grasp the concept. The flipbook is a linear sequence of pictures. At each step, the pictures change slightly the physical appearance of the original object. For example, the sequence of circles below shows the circle gradually enlarging from left to right. The visual reality sequence is the illusion of enlarging circles that result from flipping through the sequence rapidly. The faster one flips through the sequence, the smoother the transition becomes between pictures, ultimately creating the illusion of circles growing in size, or dilating.

Research has shown that a series of transformations or

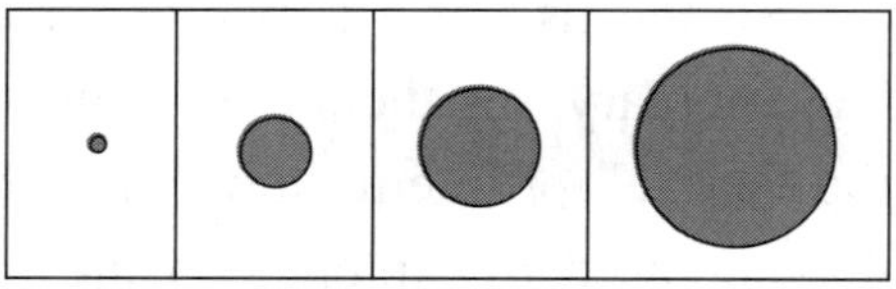

changes in a picture, or panel, can influence the thinking of young children. The concrete operational thinkers' method of coordinating mental images can lead to contradictory conclusions, because internal representations mirror earlier action sequences. The child's understanding is limited to these earlier sequences which have occurred, because she or he lacks the process of symbolic reversal of a process in order to verify and identify the inconsistencies in the action sequence. Piaget refers to the reasoning of the concrete operational child as transductive, that is, thinking from particular to particular. Therefore, the child's thinking is based on similarity rather than logical correctness.

Student teachers in math education are taught to use concrete manipulative material for teaching mathematics concepts in their required undergraduate teaching methods courses. And, as certified teachers we often find them in our supply closets. I realized their value and significance during the early stages of my career, but many teachers don't. For many teachers, manipulative devices are merely toys. I realized that the creators of these physical objects have made concrete hosts to house their understanding of mathematics. They puncture space and when manipulated in particular ways can reveal simple as well as complex mathematical ideas. These hosts serve as conceptual proxies and represent examples of pedagogical objects. The value of the manipulative devices is that they help concrete thinkers develop a deep conceptual understanding of the underlying principles of arithmetic, algebra, and geometry.[3] Successful instruction entails the procedures and methods employed to puncture space with these devices. For example, when students use geo-boards to discover that the length of the hypotenuse of the isosceles right triangle is equal to the length of a side times the square root of two, they have successfully used the "host" as a conduit towards the conceptual understanding of the concept. Students demonstrate their understanding of the concept by replaying these cognitive sequences for their teacher.[4]

[3] The devices are what I have referred to as concept proxies.

[4] Geo-boards and Dienes Blocks are examples of these recognized and widely used concepts proxies.

Chapter 5

The Geo-board as a Concept Proxy: What Are Geo-boards?

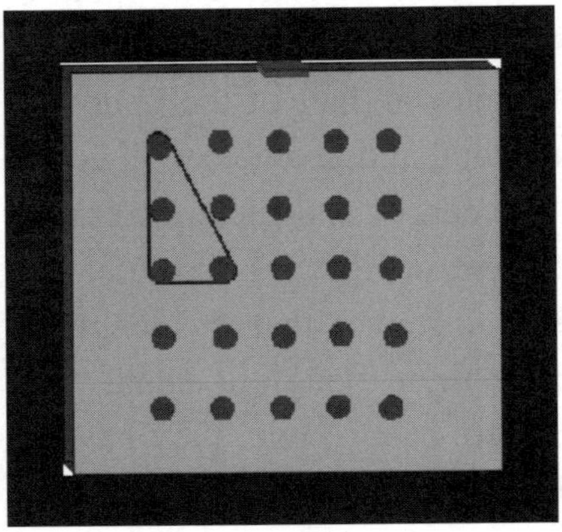

To the left of this text is a drawing of a geo-board—a rectangular lattice on which students use rubber bands to investigate geometric situations as they arise. Geometry problems can be easily demonstrated so that the study of finite geometry making use of the geo-board extends the investigation of metric space, or mensurational geometry. Since the bases can revolve and be moved about, the group of possible displacements can encompass all two-dimensional figures and properties. The geo-board has certain advantages over the chalkboard because polygonal figures created on the geo-board are variable, accurate, and clearly formed and do not depend on the skill of the teacher. With the geo-board no drawing is required; the elastic bands are simply moved from one pin to another and additional bands may be used if needed or desired. The ease with which the figures can be changed makes the tool suitably flexible.[1]

[1] Gattegno, C. For The Teaching Of Elementary Mathematics, Cuisenaire Company of America, Inc. Volume 3. 1963.

The identification and use of concept proxies has an historical basis grounded in the work of innovative individuals, such as, Caleb Gattegno and George Cuisenaire. Gattegno in particular, deserves detailed consideration and while he did extensive work in other areas,[2] I will focus here on his work in the area of geometry and the geo-board.

Caleb Gattegno of Alexandria, Egypt (1911-1988) argued that most theorems of Euclidean plane geometry could be taught using geo-boards.[3] Without knowing the specifics of Gattegno's formal education or background, I think it is fairly safe to conclude that he understood mathematics, a great deal of mathematics and that he cared deeply about others learning mathematics. His genius lay in his ability to notice, and subsequently understand, the potential of the physical geo-board to assist students learning geometry. The mere physical manifestation of the rectangular lattices that constitute the geo-board doesn't immediately suggest the fact that manipulating the lattices can make possible the discovery of geometry and geometric concepts. The geo-board allows and encourages the discovery of spatial and geometrical relationships through the manipulation of shapes and the subsequent processing and analysis of what is seen as a result of that manipulation.

[2] Gattegno did substantial work on the meaning and implementation of Cuisenaire's rods. There he saw what many had already noted but understood at once what few had been sufficiently prepared to understand. That the rods physically behaved in the way that numbers behave, and thus provided learners with an algebraic model for the study of mathematics.

[3] Gattegno, C. For The Teaching Of Elementary Mathematics, Cuisenaire Company of America, Inc. Volume 3. 1963.

Gattegno suggests:

> Geo-boards allow students to throw themselves into the activity of investigating and discover first the entire range of figures which can be formed with the aid of the rubber bands stretched over the pins, then those figures that are congruent, counting them, proving that s/he has found the right number by discovering a rational principle which allows this deduction to be made (rotation, symmetry and translation, and combinations of these); finally, s/he can discover those which are similar. S/he can set out to find all the possible quadrilaterals in the lattice, to distinguish the convex from the concave, the regular from the irregular, to classify from the general to the particular and to the still more particular. [p. 93]

Metaphorically speaking, the geo-board is a host to Gattegno's (or the teacher's) understanding of geometry, and it is through this host that he shares his understanding of geometry when he assists students in their discoveries of geometry through instructional conversations. The geo-board, like the set of chips used to teach set theory, is a concept proxy and acts as a pedagogical object when inserted into the learning environment.

The geo-boards assist logical thinkers by attaching their thinking to concrete examples, as can be described in the following example of a math lesson that is used to understand the irrational number radical two.

When I look at a geo-board I "see" how to use it to reveal my understanding of geometry—from the relationship of simple shapes to complex principles of transformational geometry. In other words, the abstraction (my understanding) is attached to an equivalent concrete representation (the board). As noted earlier, under the conceptual framework of puncturing space with pedagogical objects, instruction is defined as manipulating the elements of the objective space in ways that produce a convergence to conceptual identity between the understanding of the teacher and the students. In this specific case the elements are the concept proxy, the geo-board.

The lesson could be called: "Unit Lengths, and the Unit Length Plus a Little Bit More." It was developed to assist a sixth grade class in its study of metric areas and perimeters of polygons.

The math class had recently been investigating areas of the polygons by counting the number of square units that filled the interior of regular and irregular polygons. For example, in the diagram below the students can determine that eight square units fill the interior region of the polygon.

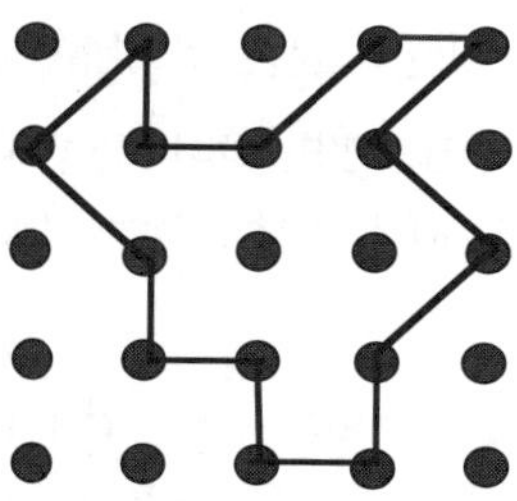

polygon

Square unit

Students recognized and commented on the fact that because of the irregular shape of the polygon several of the components of the polygon were made up of halves of the square unit. I used this opportunity to introduce the irrational number radical two. [If the question does not initiate from the students, I bring it up because it is a very important idea.] Students were asked to analyze the square unit further, giving particular attention to the effects of constructing a diagonal line between opposite vertices. Students usually observed that if the square unit has a metric

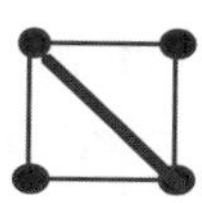

value of one, then each triangle has 1/2 the metric value of the square unit. Further analysis of the triangle leads to the realization that two sides of the triangle are congruent, that the one side is longer than the other two, and that a right angle was produced between the congruent sides. Hence, in other words the analysis leads to the observation and definition of the right isosceles triangle.

During the lesson, I helped shift the conversation to the question of how to find the perimeter of a right isosceles triangle. Students understood that each of the two shorter sides was a unit length. The discussion then focused on the diagonal side of the right triangle—the hypotenuse. The first question the students asked was "What is the length of this hypotenuse?" I asked a follow-up question, "How can we determine the length of the hypotenuse using practice methods?" In their response the class agreed that at this time we could not find the length of hypotenuse precisely but that an approximation would do for the time being.

Some students remembered an earlier debate we had had about the accuracy of measurements of the unit length using pieces of string that represented our closest approximation of the unit length. The students believed that we compare the unit length to the diagonal length and discovered and employed various ways to do this. Some used graph paper, while others used their fingers to find an acceptable approximation. Still others satisfied with the simple notion that the length was longer than that of the sides merely because it looked longer. But all of the methods employed to determine the length of the hypotenuse produced one result that everyone could accept—the hypotenuse was longer than the unit length. Determining how much longer it was than the unit length posed a very good intellectual challenge for the students. Although they could not agree on a precise value at first, they did agree that the hypotenuse was equal to a unit length and a little bit more. The process by which the students came to accept that the diagonal length was equal to the unit length + a little bit more was remarkable. On the white board in my classroom a group of students created an isosceles right triangle using masking tape and then suspended a string and black marker from the top vertex of the right triangle. The students noticed that they could construct a circle by using the vertical length of the triangle as the radius of the circle. They knew that the vertical length of the triangle was the unit length, so they know that the radius of the circle was the unit length. As they followed the path of the circumference of the newly constructed circle it became clear that the length of the diagonal of the triangle was greater than one unit length.

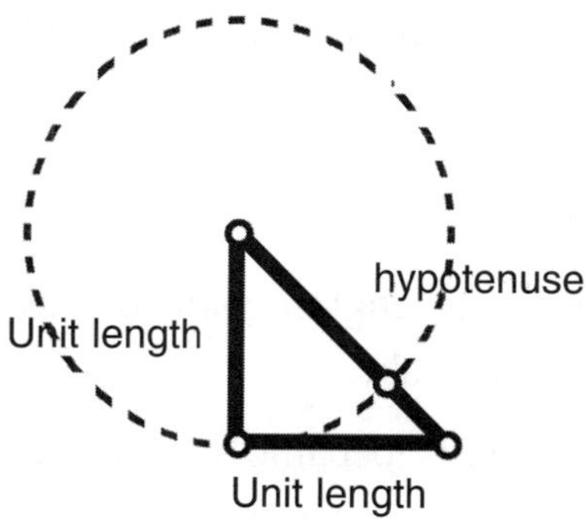

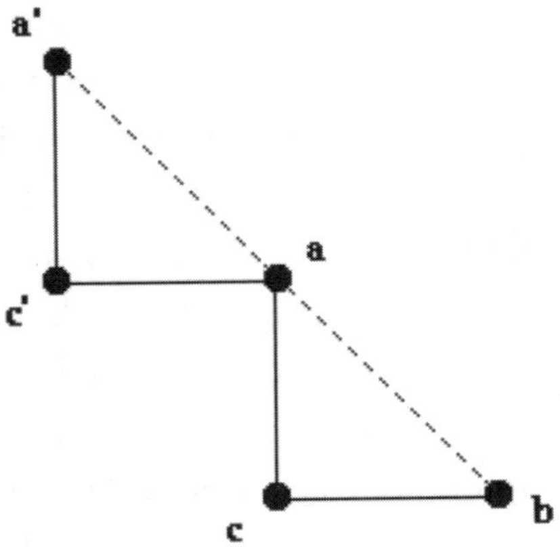

In other words, the diagonal was equal to one unit length plus "a little bit more." Symbolically, the length of the hypotenuse was equal to 1 unit length + a little bit more, or, the hypotenuse was equal to 1 + k.

The students went on to explore several other questions. For instance, how could we express two or more diagonal lengths? In the diagram below segment a'a = segment ab, and each length = 1 + k.

Student surmised that, a'b was two unit lengths plus two instances of "a little bit more." Symbolically this can be written as,

$$(1+k)+(1+k)=(1+k+1+k) = (1+1+k+k)=(1+1)+(k+k) = 2+2k.$$

Similarly three, four, n diagonal lengths could be shown to equal 3 + 3k, 4 + 4k, . . . n + nk.

But, understandably the students became curious about the numerical value of 'k', the little bit more.

Students were asked if the symbolic representation of the problem depended upon whether the value of 'a' turned out to be small, as in closer to zero, or large, as in closer to 1?

In other words, if the value of 'a' was as small as the smallest fraction they could think of, as opposed to being most of the unit length, would it change our representation of the problem? Most students recognized that it did not matter.

I sensed, however, that the students were not completely satisfied with our computations with the expression '1 + k' and wanted to know more about it. And, in particular they wanted to know just how large the "little bit more" was. The students were now ready to study the Pythagorean Theorem- $a^2 + b^2 = c^2$. At this point our main concern was put aside so that the students could develop a new mathematical tool. Students used their geo-board to understand the concept by constructing right triangles and building squares on all side of the triangle. Students then computed the areas of the square by counting the number of square units.

Since two side of the isosceles right triangle are congruent the formula becomes $a^2 + a^2 = c^2$. Substituting the values that are known into the equation we get. Then $2a^2 = c^2$ and $a^2 = c^2/2$.

The students and I then proceeded to make a series of statements (with explanations) that related the Pythagorean Theorem to our specific example and allowed us to determine the value of the "little bit more."

Statements	Reasons
(1) 12 + 12 = (1 + k)2	This is the Pythagorean Theorem. For our specific example: the sum of the squares of the two sides equals the square of the hypotenuse. In our case, the value for each side is 1.
(2) 1 + 1 = (1 + k)2	1 square is 1.
(3) 2 = (1 + k)2	
(4) $\sqrt{2} = \sqrt{(1 + k)2}$	Taking the square of both sides
(5) $\sqrt{2}$ = 1 + K	Using our hand-held calculator to compute the numerical approximation of $\sqrt{2}$ = ±1.414
(6) 1.414 = ±1+K	
(7) ±0.414 = K	subtracting 1.000 from both sides, we get ±0.414 = k

Hence the length of the hypotenuse is approximately 1.4142135623731, and the numerical approximation of "a little bit more" is 0.414. Furthermore, the perimeter of the half unit sqaure could now be represesented as 1 + 1 + w2 = ± 3.4142135623731.

Chapter 6

Combining Vygotsky and Piaget in the Classroom

Chapter III described a time in my life when I engaged in conscious and unconscious thinking about the neural basis of consciousness, thinking, learning, and teaching after an intensive reading program in these areas. Although I had not been aware of it, during that period of incubation, I was formulating a theory that produced the metaphor of teaching as "puncturing space with pedagogical objects." Several years elapsed before I formalized this metaphorical statement, but during this time I was drawn to a fascinating theoretical debate centered on the work of two significant thinkers, Lev Semenovich Vygotsky and Jean Piaget.

Jean Piaget is the most acclaimed representative of the cognitive-structuralist school. He is well known, as he is globally, for his seminal research on cognitive development. I had been introduced to his work as an undergraduate, and I subsequently used his cognitive theories as the foundation of my own work with students in my mathematics classes. However, I was less familiar with Vygotsky and his ideas, ideas that challenged Piaget's theories.

Vygotsky graduated from Moscow University with a degree in law but had interests in diverse fields—philosophy, psychology, and literature (Tharpe & Gallimore, 1988). In 1925, he completed his dissertation, "The Psychology of

Art," and thereafter he lectured, conducted research, and published in the field of psychology until his death from tuberculosis in 1934. His work was banned in the USSR for political reasons and was not published until 1956. Nevertheless, his theories have influenced a number of researchers, including, for example, Jerome Bruner. Though his original work is now more than seventy years old, it is now increasingly having an influence on educational practices in the United States (Moll, 1990).

In short, the theoretical debate between Vygotsky and Piaget centers on the genesis of thought. Piaget asserts that development occurs in distinct, measurable, and observable stages. His research and analysis focused on what children have, not what they lack. His theory assumes that development is unidirectional and that all children reach each of the stages of cognitive development at approximately the same age (Driscoll, 1994). Vygotsky (1962), on the other hand, concluded that thought and speech have distinct and different genetic roots and that no primary bond connects thought and word. They develop independently, and there is no constant correlation between them. Furthermore, he disagreed with Piaget's inference that egocentric thought is impervious to experience and with Piaget's assumption that development could not be impeded or accelerated through instruction. Vygotsky was also critical of Piaget's notion that cognitive developmental growth was based on universal characteristics. Vygotsky, however, did agree that cognitive development is complex and is affected by social and cultural contexts, although he believed that biological and cultural development are interrelated and do not develop in isolation. For Vygotsky, intellectual development was a continually evolving process without an end point (Driscoll, 1994).

What difference does all this make in how teachers practice their art? Should teachers now find any theoretical or practical significance in the theories of L. S. Vygotsky? Should their pedagogy be affected by these ideas? I would answer strongly in the affirmative.

Vygotskian ideas offer a powerful set of tools for considering how mind and consciousness develop and for directing how they can be employed in classroom practices. As such, these notions are valuable for practitioners interested in creating new kinds of educational environments and for grafting new methodologies onto their existing pedagogical practices (Rubtsov & Margolis, 1996). To my mind, Vygotsky's constructs give teachers important devices for shifting their classroom practices to a more "constructivist" approach, one that takes the specifics of context and culture into account. What are these devices? The Vygotskian perspective offers two devices that extend teaching. The first is the idea that children develop by encountering concepts or tasks that lie beyond their immediate ability to accomplish, but which are within a "zone" of possible performance that may be realized if the child works along with an adult. The second is the notion of "cultural artifacts."

These are powerful ideas and should seriously inform the practice of teaching. As Piagetians, many teachers have considered the stages of cognitive development as intellectual epochs and have acted as though they represent cognitive "gates" that define what are appropriate intellectual tasks for students at various stages of their cognitive development (e.g. pre-operational thinking versus formal thinking). These gates tend to constrain teaching by prohibiting presentation of content that is deemed inappropriate at particular stages. For instance, students at the age when they are characterized as capable of "logical thinking when it is attached to concrete examples" from the Piagetian perspective would not be taught geometry in the same manner as would older students characterized as "thinkers detached from concrete examples." As Piagetians, that is, we have believed that we must wait until children make a transition from one stage to the next before representing content at a higher level of abstract thinking. However, when the Vygotskian perspective is incorporated into our thinking about how children learn, the cognitive stages of development need not be viewed as binding. Rather it allows teachers to influence thinking at all levels with interventions targeted within what Vygotsky calls the "zone of proximal development."

Device One

One primary principle in Vygotsky's theory is what he called the zone of proximal development which he defines as the difference between children's capacity to solve problems on their own and their capacity to solve them with assistance. Children are capable of performing many functions and activities independently without the help of anyone else. But they possess, as well, a zone of proximal development that includes functions and activities that they can perform only with the assistance of someone else.

How does one know when one is, or is not teaching within the student's zone? Unfortunately, Vygotsky did not provide any useful translations of his theory into practice (Tharpe & Gallimore, 1988)—a frustrating reality for practitioners, researchers and theorists, because unless his concepts and philosophical considerations are operationally defined, they cannot be effectively dealt with in the objective space of the classroom. As I thought about how the zone of proximal development could be observed in students, I reasoned that a "trait" is never resident only in the organism, but must be seen within the context of the relationship between the organism and the environment. Vygotsky (1978) noted that in dynamic assessment, the student is helped or given instruction in order to determine the degree of independence of the student's activity. If so, when a student is unable to perform a certain activity could we regard the student as having reached the "zone of proximal development?" My thesis is that the conceptual framework of "puncturing space," and the constructs "pedagogical objects" and "concept proxies," offer a means of both observing and employing Vygotsky's notion of the zone of proximal development in the context of the classroom. Let me explain how and why I came to this conclusion by introducing the following story.

In 1990, I wrote a proposal requesting computers (concept proxy/artifact) to teach geometry to seventh and eighth graders at my present school in New York. A major component of the proposal was a statement of the underlying philosophical and theoretical considerations. I wrote that

> The Euclid Project (EP) computer-based mathematics program provides students with a study of geometry from multiple perspectives. The Euclid Project is largely based on a constructivist theoretical perspective of how mathematics is learned.
>
> According to this emerging theory of constructivist learning, students need to construct their own understanding of each mathematical concept. Hence, in addition to lecturing and explaining as a means of transferring mathematical knowledge, we believe that teaching should also create situations that will foster making the necessary mental constructions independently among students. A critical aspect of our approach is a decomposition of each mathematical concept into developmental steps following a Piagetian theory of knowledge, and a Vygotskian notion of mediated learning.

My proposal was accepted, and I was given six computers to use in my math classes. I was strongly encouraged to think creatively about content and pedagogical practices, and I was supported in my efforts by the Head Master, Assistant Head Master, and the Divisional Director of Middle School. It was a natural next step for me to think in terms of action research or inquiry. Action research became, and continues to be an important component of my "intelligent doing" because it allows me to learn meaningful things about my own teaching.

My first attempt to study my own teaching occurred during the 1993-1994 academic year. My interest in identifying when students were in their zone of proximal development prompted the following line of inquiry: I hypothesized that when students are in their zone of proximal development there will be a decrease in their on-task behavior and an increase in search behavior. Here, *on-task behavior* refers to students working at their computer stations, manipulating the elements of the provided software. *Search behavior,* on the other hand, refers to students being away from their computer stations and looking at what their peers are doing. I posed, as a working hypothesis, the proposition that students are in their zone of proximal development when the ratio of the time spent in search behavior, to time spent in on-task behavior, is greater than one.

To test this hypothesis, I eliminated myself from the learning situation as much as possible—talking with students and directing their interactions were held to a minimum—assuming that without my input into the activity, students could explore the particular features of the computer software—the Geometer's Sketchpad—freely. During this process, I made the following observations:

> Students began their exploration enthusiastically. The noise level of the classroom was low. Students quickly discovered which tools were necessary to draw a very simple closed plane figure. After a full week or so, the students began to demonstrate differing levels of technological literacy, as manifested in the quality of their explorations. Most students discovered the very simple functions of the software by pushing- or clicking buttons and observing how the computer responded. I characterized this level of discovery as 'unit-function' learning. As an example of a 'unit- function' in the Sketchpad software, when one holds down the shift button while constructing a segment, the computer will construct a straight segment. Generally, when asked what they have learned, these students responded by saying, "I know how to do this, or that." On the other hand, a small group of students began to string together several unit-functions to evoke complex responses from the software. An even smaller group of students spent time deconstructing the complex sample sketches that were developed by authors of the software. For example, some students sought to discover how the

authors animated two radii so that their endpoints rotated in opposite directions along the circumference of the circle.

At the point when students reached their potential to manipulate the software in ways that advanced their knowledge, there was an almost palpable change in the "tone" or feeling of the classroom. Students tended to talk about topics unrelated to the specific task at hand, and they moved about the classroom as if searching for novel ideas, or a new-ness. In addition, at some point, most students eventually turned to me for help, usually in the form of a request to "Show me what else I can do with the program." I rarely gave direct answers to their questions; instead, I encouraged them to change their approach.

When the tone of the classroom-setting changes in the manner that I have just described, I suggest that it is a clear indication that students are in their "zone of proximal development."

Having concluded that what I had observed was an indication of when students were in their zone, I thought about ways to exploit these moments. My story continues somewhat later in the term, during a session in which students were exploring area of polygons. Having noted that the students were in their "zone" and needed assistance, I decided to reinsert myself into the environment.

The entire class of seventeen had not reached their plateau in terms of learning something new for themselves, and therefore most students did not notice that I had re-entered the classroom nor for that matter did they seem to care about what I was doing. However, I observed, that very few students had discovered how to perform complex manipulations, such as how to animate an object or add color to the interior of the polygon using the features of the program. I decided to experiment with the construct of "puncturing space" with a new object, before the class as a whole had entered their zone. I randomly sat by two students who were busy constructing polygons, measuring their area and performing minor calculations with these numbers. I simply said to them, "Carefully watch what I am doing." At first, they did not understand the procedures for animation that I illustrated, so I repeated them two more times. The following day the majority of students in the class had at the very least heard about how to animate and colorize a polygon.

"Puncturing space" is a method for initiating and facilitating the transmission of information throughout the students' communications network. Since that first instance, I have used this strategy on numerous occasions to insert various types of information into my classes. For example, I've written highly elaborate mathematical proofs on the chalkboard to determine who would study them and raise questions. I've used this approach to determine whether students are thinking abstractly about the material they were assigned by writing statements on the chalkboard and observing which students recognize that the statements were related to the mathematics that they were studying. The most significant understanding that I have gained from these experiences of "puncturing space" is a heightened awareness that teaching is the act of inserting, deleting, and re-inserting the self into the dynamic interactions of the milieu to compete with internal and external stimuli for the attention of students. The idea, or notion, of working within the zone of proximal development of students provides the teacher with tools to look at students as they work and to determine when the time is ripe to puncture space or intervene.

Device Two

A second Vygotskian device that is useful for practice is the "artifact." The "concept proxies" that I introduced earlier are specific forms of cultural artifacts, namely pedagogical objects. Elementary level educators have a long history of using manipulative devices for the specific purpose of enhancing learning. What makes these objects artifacts? I suggest that longevity, widespread use in the working culture, and the fact that they serve a purpose are valid reasons to regard them as artifacts. Teachers mediate their activity through artifacts. For example, the math teacher uses a compass and straightedge to bisect a triangle. One of Vygotsky's major contributions was that he also viewed language and symbol systems as psychological tools for developing the human condition. Artifacts are there for us when we introduce them into a certain activity, but they are also products of our activity, and, as such, they are constantly changing through the course of activity. I have, for example, observed my daughter, at age six, use a shoehorn properly to put on her shoes after watching her mother use it for that specific purpose—and then use it as a spoon to eat her oatmeal. She had extended the use of the artifact by seeing that it was similar to her traditional spoon.

Objects, in general, do not necessarily become concept proxies without purposeful interactions between a knowledgeable individual and the recipients. A geo-board in isolation is a toy, but when combined with Gattegno's understanding of geometry, it qualifies as a pedagogical artifact because it serves as the teacher's "proxy," or "host." It can be recognized as a pedagogical artifact because it is suitable for furthering the achievement of teachers' education aims. It is not expected of students to dis-embed the comprehensive understanding of geometry, as viewed by Gattegno. However, most elementary school students have a basic understanding of shapes and can both illustrate and identify what they understand on the Geo-board. When geo-boards are inserted into space and the activities surrounding them are targeted to students' "zone of proximal educational development" they become powerful devices for enhancing their understanding of geometry.

I have argued that Gattegno's formal, abstract knowledge of geometry is equivalent to his demonstrated understanding of geometry as manifested through his manipulations of the elements of the geo-board. It is more likely that there is not a complete one-to-one correspondence between what he "knows" and his ability to demonstrate his knowledge through a concept proxy that places physical limitation on his ability to communicate the full spectrum of his knowledge. He therefore works with a sub-set of his formal knowledge of geometry. But, it remains the case that Vygotsky has offered teachers a means of interacting with artifacts to determine when the concept proxy becomes a tool for students and when it is merely a toy through observations within the activity structure.

My interest is in how artifacts that are specifically designed to host an understanding of mathematics can be used to serve a representational function. How these artifacts are invented, acquired, or transmitted across individuals or generations are also areas worthy of study but beyond the scope of this work.

It is important to emphasize that concept proxies do not simply serve to facilitate mental processes that would otherwise exist. Instead they fundamentally shape and transform them.

Human action is not always a direct response to the environment—it is mediated by culturally meaningful tools and signs that allow a person to interact with outside stimuli (control himself or herself from the outside). Collaboration with other humans creates zones of proximal development for individuals, enabling them to go beyond their current capacity by grasping and constructing new mediating tools and signs. To do a better job of unlocking human potential, we must conduct more creative and innovative experiments that improve our understanding of the conditions that induce zones of proximal development.

Speculations: The Theory of Neural Group Selection and Pedagogy

Years ago, when I attended college, I studied mathematics and the teaching of mathematics. At the graduation ceremony, the Dean of Education spoke to the graduating class about the many challenges that we would face as teachers. As I listened to his words a tiny voice in my head asked, "What have I learned about teaching?"

An important question then and now, thirty years later, I continue to ask it and struggle with the answer. The difference between my answers then and now is that many years of experience teaching mathematics are reflected in my answer of today. But, my formal undergraduate school education laid the foundation for my practice of teaching. It was then that I learned that the intellectual development of children is a progressive expansion in the structure underlying representational complexity, and that this development can be fostered by the child's interaction with the environment.

These notions of cognitive development, as formulated by Piaget, constitute my knowledge base from which I drew for explanations of variations in academic performances among my students. And, as I gained more experience teaching mathematics, his theoretical concepts became even more important. They showed me that my students' intellectual success could partially be explained by what they attend to, and the activities they engage in—and that effective learning occurs as a result of an active interaction between the learner and the environment, that learning begins, for the most part, with perception, and perception is an act of construction.

Cognitive theories like that of Piaget and other cognitive scientists have and continue to provide valuable explanations of how children learn. But, as teachers embrace the new challenges of a new millennium—of educating all our children—we must now look to the research findings and theoretical arguments from related fields, particularly those that study the brain/mind question, such as neuroscience.

The neurobiological perspective of the brain is beginning to find its way into discussions of teaching and learning.[1] I believe, for one, that it should become a major component of our thinking because important considerations for learning and implications for teaching can be drawn from this body of knowledge. Below, I present my initial thoughts, speculations, and conjectures on how theoretical neuroscience has implications for both teaching and learning.

I propose that Gerald Edelman's (1987) Theory of Neural Group Selection (TNGS) offers a valuable insight into the mind that explains the variations in academic performance among students. Of particular relevance are his conceptualizations of how "primary consciousness" develops and "re-entrant signaling." Thus, for example, I have found Edelman's theory and concepts to be helpful in my efforts to predict the behavior of my students while they are inspecting and exploring an unfamiliar visual stimulus (physical objects and symbols).

Background Information

A general notion of consciousness that refers to a state of "wakefulness" is generally accepted by most of us. But, teachers rarely, if ever, include this term in their descriptions

[1] The Master's and C.A.S. programs in Mind, Brain, and Education at the Harvard Graduate School of Education are designed for students interested in connecting cognition, neuroscience, and educational practice, especially as they

(verbal or written) of active students engaged in classroom activities. Instead, they are more likely to use such terms or phrases as “paying attention,” “attention span,” or “is selective in what the student attends to,” to refer to the classroom behavior of students that is deemed focused or unfocused. Since “selective attention” is believed to be a subset of the broader phenomena of consciousness, it made sense for me to explore what neurobiologists mean when they refer to “consciousness.”

The concept of “consciousness” poses an ambiguous problem in the science of the mind because it remains elusive. Most neuroscientists acknowledge that consciousness exists, but they do not attempt to study it, because they consider it to be a philosophical problem, best left to philosophers, or because they believe that it is premature to study it now. However, there are two groups of neurobiologists who believe that it is now time to launch a scientific attack on this phenomena.

The first group of neuroscientists believes that there are neuronal correlates of consciousness, often referred to as the NCC (Crick and Koch, 1990). They argue that at any given moment in time, it is probable that some active neuronal processes in one’s head correlate with consciousness, while others do not. They—Crick and Koch—propose that the best way to study consciousness is, first, to understand how we interpret visual information and, then, to determine where the cells responsible for determining consciousness are located. Once these cells dictating consciousness have been located, the next step would be to determine how the decisions are made about what objects we become aware of at any given point in time. The neuroscientists who subscribe to this paradigm have found the following sources of data to be useful for the analysis of NCC: (1) cases of human brain damage that alter consciousness, or studies of brain lesions in animals, (2) the use of non-invasive brain scanning methods, like functional MRI and electrophysiological recordings with the cases cited in item 1 or with animals to identify specific types of neurons with activity patterns that match their perceptual states, and (3) electrical stimulation of specific brain regions that seem to be able to shift an animal's conscious perception of a stimulus.[2]

Although this body of research is extremely interesting, the findings were difficult for me to integrate into my own classroom observations of how students acquired their knowledge. Nor was I able to fit the neuronal correlate of consciousness perspective into the Piagetian-Vygotskian conceptual framework: that knowledge acquisition is a process of continuous self-construction and that knowledge is

[2] Crick, one of the champions of this approach, contends that any object will have multiple features (form, color, motion, etc.) and therefore must be processed in different visual areas. The important question, according to Crick, is how do these cells know they are working together to produce a coherent visual perception? Crick's answer to this questions is his notion of binding. Each neuron would have certain cells attached to it which are responsible for identifying the various characteristics of objects; the binding results from when the cells attached to the firing neurons are stimulated by them based on the changes in the characteristic which they record. These specific cells know when to respond (bind) to the visual stimulus because they become the object of our attention– it is the mechanisms of attention that distinguishes one object from another in the perceptual field

invented as learners develop and interact with their surroundings.

On the other hand, Gerald Edelman's (1987) Theory of Neuronal Group Selection (TNGS) proposes that consciousness is a global brain function, implying that many widely-distributed parts of the brain are important for consciousness.

According to Edelman, consciousness results from a process that begins during the early stages of intellectual development, when the brain is busy constructing the foundation (structure) for higher order thinking. The process begins with an uncommitted population of neurons. As the child interacts with the environment (perception), it set off a competition among neurons. The neurons that win the competition do so because they are so disposed (that is evolutionary pressures have picked out specialized networks) to the stimulus, and these neurons will compete with each other in forming connections (neural networks). It is not fully known why certain neurons respond to various stimuli while others do not. Experience encourages groups of selective neurons to strengthen these connections. Gradually, these complex interactions of connections begin to stabilize, establishing the foundation of the primary host for more definitive categories. When different external stimuli, or internal mental activities are stimulated at the same time, they notify each other back and forth, thereby producing a stable neural pattern. It is this stability of neural patterns that Edelman calls thought and perception.

Edelman's construct of "primary consciousness," or maps coincide with the Piagetian stage of cognitive development referred to as sensori-motor. The neural maps develop a type of differentiation once those neurons disposed to the specific input stimuli commit themselves to the stimuli; they continue to respond and modify themselves when the stimuli are revisited through, what Edelman calls "re-entrant signaling." Re-entry occurs when an external stimulus, or an internal mental activity are simultaneously stimulated, encouraging the formation of many different neural maps, which causes millions of neurons to fire in parallel, notifying each other, back and forth and thus producing coherent patterns of neural activity. According to Edelman, it is this constant circulation of impulses that gives rise to perception and thought.

Piaget's theory of "genetic epistemology" (the study of the origins and growth of knowledge), based on the idea that knowledge acquisition is a process of continuous self-construction is compatible with Edelman's Theory of Neuronal Group Selection (TNGS). In particular, Piaget's descriptions of how young children differentiate objects in their perceptual field during the sensori-motor stages of development in terms of proximity, separation, order, and enclosure is similar to Edelman's description of the development of primary consciousness. Primary consciousness is the state of being mentally aware of objects in the world. It is where the fundamental categories—or maps—form in response to, for example, movement and the color of mental images in the present. Piaget explains the development of our initial cognitive structures as a topological represensation of objects according to the "nearby-ness" of element—the means by which objects are dissociated—how elements in the visual field are arranged one before another—where a percept is enclosed along one dimension, surrounded by another in two dimensions, or inside as in the case of an object in a box. Edelman's primary consciousness and Piaget topological representations of objects in space seem very compatible because both describe how the mind prepares itself for the development of higher-order thinking.

For Edelman, all perceptions are acts of construction and have a host of different features which could be processed in different areas of the cortex. It is, therefore, through a repeated sampling and re-entry of the elements in the external environment that one builds a representation of it. It is this continuous process of repeated sampling and re-sampling of the elements in the external environment critical to the building of a representation (a memory) of them that is important for teachers to understand, because it shows how each new or reinforced external stimulus that students confront requires them either to form new categories or to correlate existing categories, thereby establishing value and meaning. That is, the mechanisms of primary consciousness bring together into a "scene" the many categorizations involved in perception.

Now, assuming that all perception (sensory inputs) is an act of construction through the continuous process of repeated sampling and re-sampling of the elements in the external environment in order to build a representation (a memory) of it, then it seems logical that the outputs of these memories would involve a similar process. That is, the communication of thoughts (mental activity) should involve the continuous puncturing of space with parts of it (output). Although mental activity may exist in the brain as categories of neurons, it certainly does not exit the brain in this form.

What if we describe the form of mental activity when it exits in the brain (primarily through various symbols) as a combination of how we initially processed the elements that we perceived, as described by Guilford and Meeker (1969). According to them, the content (cognition, memory, divergent production, convergent production, evaluation), the form of the content (figures, words, symbols, and behavior), and the degree of complexity (as units, classes, systems, relations, transformations, implications) of the products existing in the mind are forms of the external environment—this I have called the fabrication of the Guilfordian external space. The telling and showing of mental activity will be in this form as a chain sequence of connected chunks, or bounded elements, in which each chunk in the chain sequence represents a piece of the total mental activity—the collective whole. Each smaller chunk of the chain sequence itself is a combination of content, operation, and product. This triad can be as small as a phonic symbol or as large as a physical object (a geoboard).

The chunks of the chain sequence are bounded because there are implied transitional breaks between them, as indicated by a shift from one to another. It is a well-established fact that the mind tends to categorize and organize incoming information. I suggest that one would also expect to find that exiting cognitive "chunks" have a logical order to them—a subjective order of linearity. I emphasize the concept of subjectivity because there is no certainty that what is perceived as logical, orderly, and meaningful to one individual will necessarily be perceived as such by another.

Edelman's Theory of Neuronal Group Selection offers a good explanation of why different individuals will not necessarily find the same meaning in the individual chunks or the collective chunks of the chain sequence. Edelman contends that "the job of primary consciousness is to bring together into a 'scene' the many categorizations involved in perception and that all perceptions are acts of construction." Therefore, students will either form new categories or correlate existing categories as they sample and re-sample the chunks in order for them to establish value and meaning. The union of the chunks (meanings) in the chain sequence represents learning through the processes of assimilation and accommodation. That is, students either try to fit the individual chunks into their existing cognitive structure or they modify their existing cognitive structure to fit the individual chunks.

An example of the difficulty the students may have finding value and meaning in their perception is offered by the following equation introduced by Matherne (1997) in his review of Edelman's (1989) book, *The Remembered Present*. Matherne's equation is,

Perception [now] = Function [Perception [now], Perception [past]][3]

That is, Perception [now] probably involves a lot of Perception [past] (the sum of our experiences in other contexts) and our experiences now.

Applying Matherne's equation to my reasoning, students may not find value and meaning in any of the chunks of a "chain sequence of cognitive chunks" because the chunks do not exist in the students active Perception [now]. Until students develop experiences associated with the particular chunk, thus establishing a Perception [past], which is then associated with their active Perception [now] they apparently will not relate to either the parts or the collective whole. For instance, at some point in any traditional introductory Algebra course, students will eventually confront word problems involving wind and current. Until now, before becoming familiar with Edelman's theory, I believed that success with algebraic word problems in general was a good indicator of which students would be successful in a calculus class, because these students intuitively understand the concept of simultaneity and innately practiced multi-dimensional thinking, and that students who were challenged by these types of word problems lacked an aptitude for this type of thinking.

[3] The exact function is unspecified, but may be considered as a comparison of two memory traces: one accumulated over a long period of time and one generated from the instantaneous sensory input being received. Matherne explains, the term Perception [past] to mean the integrated sum of our past perceptions. Our perception of the now is a function of our total past categorizations and our perception of the now. What we perceive now is a function of the perceptions made available to us now from our exterior surroundings (nonself) and the sum of our past perceptions stored in our interior self.

I now consider the added possibility that students are challenged by these problems because they have no related experiences of thinking in terms of simultaneous events and are perhaps not developmentally ready for multi-dimensional thinking. In other words, their active Perception [now] does not include a relevant, historical Perception [past].

Perception [now] = Function [Perception [now]].

A second observation of how TNGS appropriately predicted my students' behavior involves the acquisition of knowledge in the context of their interactions with hands-on devices designed to promote learning that are new to them (e.g., a geo-board). I've often observed that students spend a finite amount of time exploring the unfamiliar device, inspecting it from many angles repeatedly, as if searching for value and meaning. It is consistent with TNGS that students demonstrate a wide range of possible ways in their efforts to seek meaning as they explore the geo-board for the first time (no Perception [past]) to varying degrees and types of Perception [past]. My observation that students tended to select their own trajectory to inspect the geo-board is explained by TNGS which introduces the notion that different neuronal networks are forming in the brains of each student.

The fact that students did not remember what they had constructed on the geo-board a day or two before is also explained by TNGS in terms of memory. Edelman contends that what we call memory is not placed in the brain as a content, but as a re-entrant strengthening of synaptic connections, which are reinforced during the re-visit and eventually reach the level of long-term potentiation.[4]

Therefore when students ask the same questions that they had asked previously and found their own answers, or have to redo the initial inspections of the geo-board to replicate what they had discovered previously, it suggests that they require more time to construct meaning and that they need a wide variety of experiences to attain an individually acceptable level of meaning and value. It is from frequent reinforcement through re-visitation (re-entrant) of the concepts, or their design patterns that students are helped to recognize what it is that they now know—neuronal groups eventually are assigned to a task and, in the future these groups will continue to respond to the stimulus when called upon to do so, thus establishing students' Perception [now].

A second observation from my past experiences of teaching geometry to middle school students is also more easily explained by TNGS. As an undergraduate, years ago, when I was first introduced to the field of educational psychology, my professors repeatedly emphasized the importance of individual differences among students. In various classes, we discussed individual differences in students' ability to understand, apply, analyze, synthesize, and evaluate information. During my beginning years of teaching when I believed that learning was subordinate to teaching, I used the phrase "individual differences" as an euphemism for a lack

[4] Researchers have long argued that for memory and learning to occur, connected neurons must become more responsive to one another—so responsive that they continue to communicate at a high level even when they are no longer being stimulated by an external source, such as sensory input.

of "aptitude," assuming that students who could not understand mathematics the way I did after I had taught them lacked an aptitude for the subject. However, Edelman's Theory of Neural Group Selections offers an alternative explanation of individual differences that suggests that teaching should become more subordinate to learning. For instance, during the initial inspection of the geo-board, I should have anticipated that most students would be able to create intricate design patterns on the geo-board, and that very few would be able to relate their formal knowledge of mathematics to the embedded math concept in their complex design patterns because the students had no Perception [past]. I now understand that this example of the geo-board illustrates how Perception [past] affects Perception [now] because active experience "selects," or carves out, a new, more complexly- connected pattern of neuronal groups, a neuronal reflection of the students' individual experiences, of the procedures by which students come to categorize reality.

I also hypothesize that the inspection of an unfamiliar object returns students to the primary consciousness "stage" because during this stage inspections are intervals around a time chunk that Edelman calls "the present." If this is true, then one can assume that the first time that students experience an unfamiliar concept or skill, it is in terms of what they initially perceive and that they will only establish value and meaning of these experiences from frequent experiences of them.

I also now believe that students procure these necessary experiences in order for them to establish value and meaning in several ways.

To begin, learning requires a change in the learner. The activity of a teacher is relevant to the extent that it causes students to engage in activities they would not otherwise engage in—including, but not limited to, acquiring knowledge provided by the teacher, books, and through the inspection and manipulation of concrete devices that are designed to promote learning. A teacher may also engage students in tasks, some of which may involve acquisition of skills by working examples. The problem that we face as teachers is to design a series of experiences for students that will enable them to learn effectively and to motivate them to engage in the corresponding activities. The more difficult problem, and the one that often leads to different prescriptions, is determining the desirable learning goals and the experiences that, if incorporated in the instructional design, will best enable students to achieve these goals. Often, what is learned is influenced by explicit instruction because, as Simon and Feigenbaum (1964) suggest, learning is strongly influenced by the sequence of the stimuli and the feedback that tells the system when responses are correct and when they are wrong. The sequence of the stimuli is clearly compatible with my notion that mental activity exits the brain in the form of a chain sequence. In this case, my definition of teaching as the competition between the internal mental activities of my students and the external stimuli from the environment for the students' "attention" now has a neuroscience basis.

When, for whatever reason, students cannot construct the knowledge for themselves, they need some instruction. I believe that if we puncture space with pedagogical objects when students are in their zones of proximal development (Vygotsky, 1978), students will both assimilate and accommodate the concept or skill in accordance with Piagetian theory. Teachers must also consider whether or not students have an active Perception [now], which includes a valid Perception [past].

I believe that students are in their "zone" when they demonstrate off-task behavior during small group work, or are unable to respond to questions during lectures. Indications of when students are off-task is observed when, for example, the focus of their internal mental activity bears no relationship to the topic at hand (e.g., talking about a movie), or when I sense that they are unable to respond to my questions. These off-task behaviors could indicate that they have not made the jump to the next level in their thinking.

When I replace Matherne's equation of Perception [now] with internal mental activities and external stimuli, then student learning = F(Perception [now] - student, Perception [now]– teacher). The activity of a teacher is relevant to the extent that it causes students to engage in activities they would not otherwise engage in. When students cannot construct the knowledge for themselves, they need some help (instruction).

A Final Note

Dan C. Lortie's (1973) celebrated and highly respected text, *Schoolteacher: A Sociological Study* has significantly influenced the educational establishment, both at the local and federal levels, and has brought clarity to and greater understanding of the importance of "culture" of schools.

Educators have long recognized a simple reality, that education occurs behind closed doors, in isolation, in classrooms where teachers have the greatest impact on what students learn, and how they learn; that it is behind closed doors where teaching occurs, teaching styles develop, and beliefs about the learning process are implemented. Teachers shape and mold their classroom environments in ways that confirm their pedagogical beliefs and practices. It is also behind closed doors where classroom teachers quickly learn that they must develop the ability to process multiple aspects of their environment simultaneously and manage conflicting roles. In time, teachers develop their professional "voice," a complex product of their experiences in the roles of educational theorist, psychologist, philosopher, curriculum developer, researcher, and human being.

I believe that all voices must be allowed to flourish in the public arena, where they can be evaluated and reviewed by peers. In time, and with consistent refinement, their descriptions of the practice will develop into a useful form, a form that will satisfy those concerned with theory development, and research. Until teachers learn what is essential in order for them to share their individual theories of practice with those who specialize in theory development and research, I agree with Paul F. Brandwein (1966) when he states that,

> We must allow an investigator into the elements of the teaching art at least to state his abstraction, or his hypotheses, to his own satisfaction. As we learn more and more from further observation, the ambiguous theory is clarified, and the operations which the theory itself directs may be administered in self-correction. (p. 2)

This book is my abstraction. I wrote this book to demonstrate that "those who can, also teach." But, as I wrote earlier, the initial purpose transformed into a new and, perhaps more informative purpose. As I told my stories, I began to listen to my inner voice speak with a new vocabulary, with phrases like "puncturing space," "pedagogical objects," and "concept proxy" gaining new meaning from their direct attachment to what I do—teach. They belong to the contents of my mental activity that guide my act of teaching.

A newly found world of relevant information was revealed to me as I searched for clarity and validity in my own thinking. It is one thing to state my abstraction, it is another thing to find support for it.

I hope that *Teaching Mathematics: Puncturing Space: A Developing Pedagogical Tool* is a contribution to the field, because it truly represents the development of theory in my attempts to both interpret important ideas related to learning, synthesizes these ideas, and translates them into practical applications.

Bibliography

Adolphs R, Tranel D, Damasio H, Damasio A (1994) Impaired recognition of emotion in facial expressions following bilateral damage to the human amygdala. Nature 372: 669-672.

Alesandrini, K. (1981). Pictorial-verbal and analytic-holistic learning strategies in science learning. Journal of Educational Psychology, 73(3), 358-368.

Behr, M. J., & Eastman, P.M. Interactions between structure-of-intellect factors and two methods of presenting concepts of modulus seven arithmetic—A follow-up and refinement study. Journal for Research in Mathematics Education, 1975, 6, 150-157.

Blake R 1989 A neural theory of binocular rivalry. Psychol. Review 96: 145-167

Blais, DM (November 1988). "Constructivism: A Theoretical Revolution for Algebra." Mathematics Teacher 624-631.

Bredderman, T.A. The effects of training on the development of the ability to control variables. Journal of Research in Science Teaching, 1973, 10(3), 189-200.

Bredderman, T. A. Elementary school science experience and the ability to combine and control variables. Science Education, 1974, 58(4), 457-469.

Brophy, Jere. ON MOTIVATING STUDENTS. Occasional Paper No. 101. East Lansing, Michigan: Institute for Research on Teaching, Michigan State University, October 1986. 73 pages. ED 276 724.

.... "Synthesis of Research on Strategies for Motivating Students To Learn." EDUCATIONAL LEADERSHIP (October 1987): 40-48. EJ 362 226.

Brooks, J. Grennon, & Brooks, M. G. In Search of Understanding: The Case for Constructivist Classrooms. Association for Supervision and Curriculum Development, 1993.

Brooks, M.G., and J.G. Brooks. (Fall 1987) "Becoming a Teacher for Thinking: Constructivism, Change, and Consequence." The Journal of Staff Development 8,3: 16 - 20

Bishop, J. "Can There Be Thought Without Language?" In Thinking: The Expanding Frontier. Edited by W. Maxwell. Philadelphia: Franklin Institute, 1983.

Brandwein, Paul F. "Notes Towards a General Theory of Teaching", New York: Harcourt, Brace & World, 1966.

Bruner, J.S. "Language as an Instrument of Thought." In Problems of Language and Learning. Edited by A. Davis. London: Heinemann, 1975.

Campbell, D. T., & Stanley, J.C. Experimental and quasi- experimental designs for research on teaching. In Gage (Ed.) Handbook of Research on Teaching. Chicago: Rand McNally, 1963.

Carter, K. (1993). The place of story in the study of teaching and teacher education. Educational Researcher, 22,1: 5-12.

Case, R. Structures and Strictures: Some functional limitations on the course of cognitive growth. Cognitive Psychology, 1974, 6 544-573.

Case, R. Gearing the demands of instruction to the developmental capacities of the learner. Review of Educational Research, 1975, 45, 59-87.

Cattell, R.B. Abilities: Their structure, growth, and action. Boston: Houghton Mifflin, 1971.

Chalmers D 1995 The Conscious Mind: In Search of a Fundamental Theory. Oxford University Press, Oxford

Clandinin, D. J., & Connelly, F. M. (1995). Teachers' professional knowledge landscapes. New York: Teachers College Press.

Clandinin, D. J., Davies, A., Hogan, P., & Kennard, B. (1993). Learning to teach, teaching to learn: Stories of collaboration in teacher education. New York: Teachers College Press.

Clark C. M and Yinger, R. J "Three Studies of Teacher Planning," East Lansing, Mich.: Institute for Research on Teaching, June 1979.

Cobb, P. (1990). A constructivist perspective on information-processing theories of mathematical activity. International Journal of Educational Research, 14, 67-92.

Cobb, P., Yackel, E. & Wood, T. (1992). A constructivist alternative to the representational view of mind in mathematics education. Journal for Research in Mathematics Education, 23, 2-33.

Cochran-Smith. M. & Lytle, S. (1990). Research on teaching and teacher research: The issues that divide. Educational Researcher, 19, 2:2-11.

Cochran-Smith, M., & Lytle, S. (1993). Inside outside: Teacher research and knowledge. NY: Teachers College Press.

Coles, R. (1989). The call of stories. NY: Houghton Mifflin.

Condry, J., and J. Chambers. "Intrinsic Motivation and the Process of Learning. In THE HIDDEN COSTS OF REWARD, edited by M.R. Lepper and D. Greene. 61-84. Hillsdale, New Jersey: Lawrence Erlbaum Associates, Inc., 1978.

Confrey, J. (1995). How compatible are radical constructivism, sociocultural approaches, and social constructivism. In J. Gale (Ed.), Constructivism and Education. Broadway, Hillsdale: Lawrence Erlbaum Associates, Publishers.

D'Ambrosio, U. (2001). What is ethnomathematics, and how can it help children in schools. Teaching Children Mathematics, 7(6), 308-.

Connelly, F. M., & Clandinin, D. J. (1990). Stories of experience and narrative inquiry. Educational Researcher, 19, 5: 2-14.

Crick F C, Koch C 1998 Consciousness and neuroscience. Cerebral Cortex 8: 97-107

Dienes, Zoltan P., and E. W. Golding. Geometry through Transformation. Vols. 1-3. New York: Herder & Herder, 1967.

diSessa, A. (1979). Learnable representations of knowledge. In J. Lockheed & J. Clement (eds.), Cognitive process instruction. Philadelphia: Franklin Institute Press.

diSessa, A. (1985). Learning about knowing. In E. Klein (Ed.), Children and computers: New directions for child development (28), (pp. 97-124). San Francisco: Jossey-Bass.

diSessa, A. (1988) Knowledge In Pieces. In G. Forman & P. Pufall (Eds.) Constructivism in the Computer Age (pp. 49 -70). Hillsdale: Lawrence Erlbaum Associates.

Dillashaw, F.G., & Okey, J. R. (1980). Test of the integrated science process skills for secondary science students. Science Education, 64(5), 601-608.

Doyle, W. "The Uses of Nonverbal Behaviors: Toward an Ecological Model of Classrooms," Merrill-Palmer Quarterly, July 1977.

Dreyfus, Tommy & Hadas, Nurit: "Euclid May Stay-and Even Be Taught, In Learning and Teaching Geometry", edited by Mary Montgomery Lindquist and Albert P. Shulte, pp. 47-58. Reston, Va.: National Council of Teachers of Mathematics, 1987.

Driscoll, Marcy P. (1994). Psychology of learning for instruction. Boston, MA: Allyn and Bacon.

Eastman, P. M. & Behr, M.J. Interaction between structure of intellect factors and two methods of presenting concepts of logic. Journal for Research in Mathematics Education, 1977, 8, 379-381.

Eastman, P. M. & Carry, L.R. Interaction of spatial visualization and general reasoning abilities with instructional treatment in quadratic inequalities: A further investigation. Journal for Research in Mathematics Education, 1975, 6, 142-149.

Edelman, G.: 1987. Neural Darwinism, New York: Basic Books.

Eisner, E. W. "Education Objectives: Help or Hindrance? "School Review 75, No. 3 (1967) pp. 250-266.

Elbaz, F. (1983). Teaching thinking: A study of practical knowledge. New York: Nichols. Florio-Ruane, S. (1991). Conversation and narrative in collaborative research. In C. Witherell, & N. Noddings (Eds.), Stories lives tell: Narrative and dialogue in education. NY: Teachers College Press. 234-256.

Epstein, H. T. "Phrenoblysis: Special Brain and Mind Growth Periods: Human Brain and Skull Development," Developmental Psychology, 7 , 1974, pp. 207-216.

Epstein, H.T. "Growth Spurts During Brain Development Implications for Education Policy and Practice," in Education and the Brain: National Society for the Study of Education (79th Yearbook, Part II). Edited by J. Chall. Chicago: University of Chicago Press. 1978, pp. 343-370.

Epstein, H. T. "Correlated Brain and Intelligence Development in Humans," in Development and Evaluation of Brain Size: Behavioral Implications. Edited by M. Hahn. New York: Academic Press, 1979a.

Epstein, H.T. "Some Biological Bases of Cognitive Development." Bulletin of the Orton Society (1980): 30.

Flavell, J. H., The Developmental Psychology of Jean Piaget, London: Van Nostrand, 1963.

Forman, G., and P.B. Pufall, eds. Constructivism in the Computer Age. Hillsdale, N.J.: Lawrence Erlbaum Associates.

Gage, N. & Berliner, D. (1992). Educational psychology (5th ed.), Princeton, New Jersey: Houghton Mifflin Company.

Gage, N.L. The Scientific Basis of the Art of Teaching, New York: Teachers College Press, 1978.

Gazzaniga, M. S.: 1985. The Social Brain, New York: Basic Books.

Ginsburg, H.P. (1981). Piaget and education: the contributions and limits of genetic epistemology. In I.E. Sigel, D. M. Brodzinsky, & R. M Golinkoff (eds.) New directions in Piagetian theory and practice (pp. 315-330). Hillsdale, Lawrence Erlbaum Associates.

Gordon, W.J.J., Synectics. New York, NY: Harper and Row, 1961.

Glasersfeld, E. von (1990). "Environment and Education." In L.P. Steffe & T. Wood (eds.), Transforming Children's Mathematics Education: International Perspectives, (pp. 200-215). Hillsdale, NJ: Lawrence Erlbaum.

Glasersfeld, E. von (1992). "Questions and Answers About Radical Constructivism." In M.K. Pearsall (ed.), Scope, Sequence, and Coordination of Secondary Schools Science, Vol. 11, Relevant Research, (pp. 169-182). Washington DC: NSTA.

Guilford, J. P. The Nature of Human Intelligence, New York: McGraw-Hill Book Company, 1967.

Guilford, J.P. (1950). Creativity. American Psychologist, 5, 444-454.

Guilford, J.P. & Hoepfner, R. (1971). The Analysis of Intelligence. New York: McGraw-Hill.

Guilford, J.P. (1982). Cognitive psychology's ambiguities: Some suggested remedies.

Pyschological Review, 89, 48-59. Meeker, M.N. (1969). The Structure of Intellect. Columbus, OH: Merrill.

Hollingsworth, S., Cody, A., Davis-Smallwood, J., Dybdahl, M., Gallagher, P., Gallego, M., Maestre, T., Minarik, L. T., Raffel, L., Standerford, N. S., & Teel, K. M. (1994). Teacher research and urban literacy education. New York: Teachers College Press.

Hopfinger, J. B., Buonocore, M. H., and Mangun, G. R. (2000). The neural mechanisms of topdown attentional control. Nature Neuroscience 3. 3: 284 .

Hudgins, Bryce. A Catalogue of Concepts in the Pedagogical Domain of Teacher Education, Washington Univ., Grad. Institute of Education, St. Louis, Mo., 1974.

Huff, D., and I. Greise. How to Take a Chance. New York: W.W. Norton & C0., 1959.

Inhelder, B., & Piaget, J. The growth of logical thinking from childhood to adolescence. New York: Basic Books, 1958.

Izaak Wirszup and Robert Streit, pp. 233-34. Reston, Va.: National Council of Teachers of Mathematics, 1987.

Kreiman, Gabriel, Itzhak Fried, and Christof Kock (2000). Category-specific visual responses of single neurons in the human medial temporal lobe. Nature Neuroscience 3. 9: 946 - 953.

Kreiman, Gabriel, Itzhak Fried, and Christof Kock (2000). Imagery neurons in the human brain. Nature 408 .6810: 357 - 361.

Kreiman, G., Fried, I. and Koch, C. (2002) Single neuron correlates of subjective vision in the human medial temporal lobe. Proceedings of the National Academy of Sciences USA, 99:8378-8383.

Landis, B. "Ego Boundaries," Psychological Issue Monograph 24 (1970) 6.

Lawson, A. E. Formal operations and field independence in a heterogeneous sample. Perceptual and Motor Skills, 1976, 42, 981-982.

Lawson, A. E.., & Wollman, W. T. Encouraging the transition from concrete to formal cognitive functioning: an experiment. Journal of Research in Science Teaching, 1976, 13(3) 413-430.

Lesh, R. & Lamon, S. J. (1992). Assessment of authentic performance in school mathematics. Washington, DC: AAAS Press.

Lepper, Mark R. "Motivational Considerations in the Study of Instruction." COGNITION AND INSTRUCTION 5, 4 (1988): 289-309.

Lewis, A.C. "Reading the Writing on the Wall." Phi Delta Kappan 69, no. 7 (1988): 468-69.

Ling, Jon. " SMP 11-16: The Most Recent Work in Curriculum Development by the School Mathematics Project, and Its Relation to Current Issues in Mathematical Education in England." In Developments in School Mathematics around the World, edited by Meyers, C., and T. Sallee. Make It Simpler, a Practical Guide to Problem Solving in Mathematics, Menlo Park, Calif.: Addison-Wesley Publishing Co., 1983.

Lortie, Dan . Schoolteacher: A Sociological Study, Chicago: University of Chicago Press, [1975]

Logothetis N 1998 Single units and conscious vision. Philosophical Transactions of the Royal Society of London B, 353: 1801-1818

Miura, I. T. (2001). The influence of Language on Mathematical Representations. In F. R. Curcio (Ed.), The roles of representation in school mathematics: 2001 yearbook. Reston: National Council of teachers of Mathematics.

Moll, L. (Ed.) (1990). Vygotsky and education: Instructional implications and applications of sociohistorical psychology. Cambridge, MA: Cambridge University Press.

Linn, M.C. Influence of cognitive style and training on tasks requiring the separation of variables schema. Child Development, 1978, 49 874-877.

Lochhead, J. (1988). "Some Pieces of the Puzzle." In Constructivism in the Computer Age, edited by G. Forman and P. Pufall. Hillsdale, N.J.: Lawrence Erlbaum Associates.

Lozanov, G. Suggestology and Outlines of Suggestoedy, New York: Gordon and Breach, 1978.

MacDonald, J. B. "Myths about Instruction," Educational Policy 22, No. 8 (1967) pp. 571-576, 609-617.

March, J. G., & Simon, H. A. Organizations, New York: John Wiley & Sons, Inc., 1958.

McCutcheon, G. "How Do Elementary School Teachers Plan? The Nature of Planning and Influences on It," University of Virginia, February 1979.

Moll, L. (Ed.) (1990). Vygotsky and education: Instructional implications and applications of sociohistorical psychology. Cambridge, MA: Cambridge University Press.

Neimark, E. D. Towards the disembedding of formal operation from confounding with cognitive style. Paper presented at the meeting of the Piaget Society, Philadelphia, May 1977. Report No. 25, Douglas College, State University of New Jersey.

Noddings, H (1990). "Constructivism in Mathematics Education." Journal of Research in Mathematics Education #4. Reston ,Va.: NCTM.

Olson. and H.W. Stevenson. Hillsdale, N.J.: Lawrence Erlbaum Associates.

Office of Technology Assessment. Technology and the American Transition. Washington, D.C.: Government Printing Office, 1988.

Paivio, A.: 1986. Mental representations: A dual-coding approach. Oxford: OUP.

Paley, V. (1979). White teacher. Cambridge MA: Harvard University Press.

Pasch, M. (1993). Staff development and student achievement: Making promises you might have to keep. Journal of Staff Development 14,3, 13-14.

Papert, S. (1988). "The Conservation of Piaget: The Computer as Grist to the Constructivist Mill." In Constructivism in the Computer Age, edited by G. Forman and P.B. Pufall. Hillsdale, N.J.: Lawrence Erlbaum Associates.

Peterson, P. L. & Clark, C.M. "Teachers Reports of their Cognitive Processes During Teaching," American Educational Research Journal 15, No. 4 (Fall 1978) pp. 555-565.

Piaget, J. The Origins of Intelligence in Children, New York: International Universities Press, 1952.

Piaget, J. The Child's Conception of Space, New York: W. W. Norton & Company, Inc, 1967.

Polkinghorne, D. (1988). Narrative knowing in the human sciences. Albany: SUNY

Reynolds, Chelazzi and Desimone,1999, The Journal of Neuroscience, 19(5):1736-1753

Reys, B. J. (1984). Affective variables and mathematics education. Elementary School Journal, 84, 558-581.

Pollak, Henry. Notes from a talk given at the Mathematical Sciences Education Board. Frameworks Conference, May 1987, at Minneapolis.

Rogoff, B. (1990). "Apprenticeship in thinking: Cognitive development in social context." Oxford: Oxford University Press.

Rubtsov, V., & Margolis, A. (1996). Activity-oriented models of information-based instructional environments. (1996.) In S. T. Kerr (Ed.) Technology and the future of schooling: A critical perspective (pp. 172-199). 95th NSSE Yearbook, Part II. Chicago: University of Chicago Press.

Siegler, R.S., Liebert, R. M. Acquisition of formal scientific reasoning by 10 & 13 year olds: Redesigning a factorial experiment. Developmental Psychology, 1975, 11, 401-402.

Schön,D.1983 The Reflective Practitioner.Basic Books New York.

Sternberg, R. J. Criteria for Intellectual Skills Training," Educational Research 12 (2),6-12, 1983.

Taylor, Philip H. How Teachers Plan Their Courses, New York: National Foundation for Education Research, 1970.

Tharpe, R. G. and Gallimore, R. (1988). Rousing minds to life. Cambridge, MA. Cambridge University Press.

Turkle, S. (1984). The second self: Computers and the human spirit. New York: Simon & Shuster.

Tyler R.W. Basic Principles of Curriculum and Instruction, Chicago: University of Chicago Press, 1950.

Scandura, Joseph M., Mathematics: Concrete Behavioral Foundations, New York, Evanston, San Francisco, London: Harper & Row, 1971.

Scardamalia, M., and C. Bereiter. (1983). "Child as Coinvestigator: Helping Children Gain Insight into Their Own Mental Processes. In Learning and Motivation in the Classroom, edited by S.G. Paris, G.M.

Sigel, I. "The Piagetian System and the World of Education," in Studies in Cognitive Development: Essays in Honor of Jean Piaget. Edited by D. Elkind and J. H. Flavell. New York: Oxford Press. 1969, pp. 465-489.

Sigel, I.E. (1978). "Constructivism and Teacher Education." The Elementary School Journal 78,5.

Simon, H. A. & Feigenbaum, E. A. (1964). In information-processing theory of some effects of similarity, familiarization, meaningfulness in verbal learning. Journal of Verbal Learning and Verbal Behavior, 3, 385-396.

Vygotsky, L.S. (1978). "Mind in society: The development of higher psychological processes" (M. Cole, V. John-Steiner, S. Scribner, & E. Souberman, Eds. and Trans.). Cambridge, MA: Harvard University Press.

von Glasersfeld, E. (1981). "The Concepts of Adaptation and Viability in a Radical Constructivist Theory of Knowledge." In New Directions in Piagetian Theory and Practice, edited by I.E. Sigel, Brodinsky, and Golinkoff. Hillsdale, N.J.: Lawrence Erlbaum Associates.

Wertsch, J. V. (1991) Voices of the mind: A sociocultural approach to mediated action. Cambridge, MA: Harvard University Press.

Wertsch, J. (1985). "The Concept of Internalization in Vygotsky's Account of the Genesis of Higher Mental Function." In Culture, Communication and Cognition. edited by J Wertsch. Cambridged: Cambridged University Press, pp. 162-182.

Whitehead,J. 1972 A preliminary investigation of the process through which adolescents acquire scientific understanding. Unpub. M.A. dissertation. University of London.

Whitehead,J. 1985 An Analysis of an Individual's Educational Development: The Basis for Personally Oriented Action Research. Chapt. in Educational Research: Principles, Policies and Practices. Ed.Shipman, M. Falmer. London.

Whitehead, J. & Foster, D 1984 Action Research and professional educational development. Cambridge Action Research Network. Bulletin No.6, pp41-45.

Whitehead, J., "Creating A Living Educational Theory From Questions of the Kind, 'How Do I Improve My Practice?' Cambridge Journal of Education, Vol. 19, No. 1, 1989.

Witherell, C., & Noddings, N. (Eds.). (1991). Stories lives tell: Narrative and dialogue in education. New York: Teachers College Press.

Witkin, H. A., Moore, C.A., Goodenough, D. R. & Cox, P.W. Field-dependent and field-independent cognitive styles and their educational implications. Review of Educational Research, 1977, 47, 1-64.

Witkin, H. A., Oltman, P. K., Raskin, E., & Karp, S. A. A manual for the embedded figures tests. Palo Alto: Consulting Psychologists Press, 1971.

Yinger, RJ. Study of Teacher Planning: Description and Model of Preactive Decision Making, East Lansing, Mich., Inst. for Research on Teaching, July 1978.

Zahorik, J.A., "Teachers' Planning Models," Educational Leadership, 33, 1975 pp. 134-139.

Zahorik, J. A., "The Effect of Planning on Teaching," Elementary School Journal 71 (1970), pp. 143-151.

Driscoll, Marcy P. (1994). Psychology of learning for instruction. Boston, MA: Allyn and Bacon.

Moll, L. (Ed.) (1990). Vygotsky and education: Instructional implications and applications of sociohistorical psychology. Cambridge, MA: Cambridge University Press.

Tharpe, R. G. and Gallimore, R. (1988). Rousing minds to life. Cambridge, MA. Cambridge University Press.

Valsiner, J. (1993) Culture and human development: A co-constructivist perspective. In P. Van Geert & L. Moss (Eds.), Annals of theoretical psychology, Vol. X. New York: Plenum.

Vera, A. H. & Simon, H. A. (1993). Situated action: A symbolic interpretation. Cognitive Science, 17, 7-48

Vygotsky, L. (1962). Thought and language. Cambridge, MA. MIT Press.

Vygotsky, L. (1978). Mind in Society: The Development of Higher Psychological Processes. Cambridge, MA: Harvard University Press.

Wozniak, R.H. (1993) Co-constructive metatheory for psychology: Implications for an analysis of families as specific social contexts for development. In. R.H. Wozniak & K.W. Fischer (Eds.), Development in context: Acting and thinking in specific environments. Hillsdale, NJ: Erlbaum.